De Minimis Risk

CONTEMPORARY ISSUES IN RISK ANALYSIS
Sponsored by the Society for Risk Analysis

De Minimis Risk

Edited by
Chris Whipple

Energy Study Center
Electric Power Research Institute
Palo Alto, California

Plenum Press • New York and London

Library of Congress Cataloging in Publication Data

De minimis risk / edited by Chris Whipple.
      p. cm. — (Contemporary issues in risk analysis; v. 2)
  Proceedings of a workshop sponsored by the Society for Risk Analysis and others, held May 29–30, 1985, at the Brookings Institution in Washington, D.C.
  Includes bibliographical references and index.
  ISBN-13: 978-1-4684-5295-2   e-ISBN-13: 978-1-4684-5293-8
  DOI: 10.1007/978-1-4684-5293-8
  1. Risk assessment — Congresses. I. Whipple, Chris G. II. Society for Risk Analysis. III. Series.
T174.5.D4 1987
658 — dc19   87-27590
                                               CIP

© 1987 Plenum Press, New York
Softcover reprint of the hardcover 1st edition 1987
A Division of Plenum Publishing Corporation
233 Spring Street, New York, N.Y. 10013

Contributors

Daniel Byrd III   Halogenated Solvents Industry Alliance, Washington, D.C. 20036

Cyril Comar†   Cornell University, Ithaca, New York 14853; Environmental Assessment Department, Electric Power Research Institute, Palo Alto, California 94303

Joyce P. Davis   General Physics Corporation, Columbia, Maryland 21044

Raymond E. Donnelly   Occupational Safety and Health Administration, U.S. Department of Labor, Washington, D.C. 20210

Joseph Fiksel   Teknowledge, Palo Alto, California 94303

W. Gary Flamm   Center for Food Safety and Applied Nutrition, Food and Drug Administration, Department of Health and Human Services, Washington, D.C. 20204

R. Scott Frey   Department of Sociology, Anthropology, and Social Work, Kansas State University, Manhattan, Kansas 66506

L. Robert Lake   Center for Food Safety and Applied Nutrition, Food and Drug Administration, Department of Health and Human Services, Washington, D.C. 20204

Lester Lave   Graduate School of Industrial Administration, Carnegie-Mellon University, Pittsburgh, Pennsylvania 15213

Ronald J. Lorentzen   Center for Food Safety and Applied Nutrition, Food and Drug Administration, Department of Health and Human Services, Washington, D.C. 20204

Charles B. Meinhold   Safety and Environmental Protection Division, Brookhaven National Laboratory, Upton, New York 11973

Joshua Menkes   Lagos, Portugal

Sheldon Meyers   Office of Radiation Programs, Office of Air and Radiation, Environmental Protection Agency, Washington, D.C. 20460

† Deceased.

Paul Milvy   Environmental Protection Agency (WH550 D), Washington, D.C. 20460

Samantha A. Richter   Health and Safety Research Division, Oak Ridge National Laboratory, Oak Ridge, Tennessee 37830

Alan M. Rulis   Center for Food Safety and Applied Nutrition, Food and Drug Administration, Department of Health and Human Services, Washington, D.C. 20204

Patricia S. Schwartz   Center for Food Safety and Applied Nutrition, Food and Drug Administration, Department of Health and Human Services, Washington, D.C. 20204

Miller B. Spangler   Office of Nuclear Reactor Regulation, Nuclear Regulatory Commission, Washington, D.C. 20555

Curtis C. Travis   Health and Safety Research Division, Oak Ridge National Laboratory, Oak Ridge, Tennessee 37830

Terry C. Troxell   Center for Food Safety and Applied Nutrition, Food and Drug Administration, Department of Health and Human Services, Washington, D.C. 20204

Alvin M. Weinberg   Institute for Energy Analysis, Oak Ridge Associated Universities, Oak Ridge, Tennessee 37830

Chris Whipple   Energy Study Center, Electric Power Research Institute, Palo Alto, California 94303

Preface

On May 29 and 30, 1985, a workshop was held to explore the legal, ethical, social, scientific, and practical aspects of the use of the de minimis risk concept for health and safety regulation. The workshop was sponsored by the Society for Risk Analysis and its National Capital Area Chapter, the Environmental Protection Agency, the Nuclear Regulatory Commission, and the Electric Power Research Institute. The two-day meeting was held in Washington, D.C., at the Brookings Institution; however, the Brookings Institution was not a sponsor of the meeting and did not play a role in its program.

De minimis risk policy considerations were addressed from a theoretical and philosophical viewpoint, from a quantitative and methodological basis, and through insights gained with regulatory applications. The distinctions between these three approaches to the subject are not sharp; most papers in these proceedings address aspects of all three topics.

The reader familiar with the literature on the use of risk assessment in regulatory policy and decision making will find significant new contributions to the field. One of these is the examination of regulatory actions—in particular actions by the EPA—in response to risks of varying magnitude. Many attempts to seek patterns in regulatory policies have been based on analysis of the implicit economic value in obtaining risk reductions. These analyses have typically found great variability in the marginal cost-effectiveness of regulatory actions.[1] Several papers here examine regulatory actions associated with a large number of chemical carcinogens; from this starting point, clear patterns emerge when both risk levels and the size of populations at risk are considered.

A second significant contribution to the risk management literature is the collection in one volume of detailed studies of the history and current use of de minimis policies in radiation protection. In addition to the institutional perspectives of the EPA and the National Commission on Radiation Protection and Measurements provided in the chapters by Sheldon Meyers and Charles Meinhold, Joyce Davis and Miller Spangler provide extensive details of how the de minimis concept has been suggested and used for radiation. Many issues now emerging in other regulatory contexts have useful parallels in radiation.

---

1. John Graham and James Vaupel, "The Value of a Life: What Difference Does it Make?" *Risk Analysis*, 1(1) March 1981, 89–95.

For example, the distinction between, and consequences of, management of risks to individuals versus risks to populations has been recognized for many years in the radiation protection community. The use of a de minimis concept for the classification of materials as radioactive or not is equivalent to the distinction between hazardous and nonhazardous wastes in other fields.

The volume opens with chapters which weigh the attractions and drawbacks of adopting a de minimis policy in risk regulation, and which consider the foreseeable policy questions such implementation will entail. I know of no clearer introduction to these issues and explanation of the argument for a de minimis policy than that written by my late colleague, Cyril Comar, in an editorial published in *Science* in 1979, reprinted here with permission from the AAAS. The manner in which this essay approaches risk management and de minimis risk is remarkably fresh and insightful. Joseph Fiksel succinctly describes the de minimis risk concept and summarizes the conclusions of a recent study of the subject he coauthored for the National Science Foundation. Fiksel's first conclusion—that designation of a risk as de minimis is meaningful only in the context of the activity that produces it and to the party that bears it—was an unchallenged consensus of the workshop. Joshua Menkes and R. Scott Frey take a less optimistic view of the prospects for positive contributions to risk management from de minimis, for two reasons. First, they ask, "Who has the authority to define specific numerical de minimis guidelines?" Their second concern is with the accumulation of many small risks into a significant total. My chapter argues that a de minimis risk approach to small risks has attractions to regulators, the regulated community, and the scientific community. However, several potential obstacles to the widespread implementation of a de minimis policy in regulation are noted; it is argued that public acceptance of de minimis may be the most important factor in whether it can be applied.

Alvin Weinberg's chapter provides a bridge from policy issues to scientific ones. His discussion, also presented to a National Academy of Engineering Symposium held about the same time as this workshop, defines the regulator's dilemma as the requirement by law to regulate risks that current science cannot measure. The nature of uncertainty in different types of risks is described, with discussion of which uncertainties appear irreducible from our current perspective, as well as descriptions of risks with promising prospects for reductions in uncertainty. Dr. Weinberg describes how the regulator's dilemma threatens the credibility of scientists and of science itself. As a partial solution, use of a de minimis policy for small risks where uncertainties are greatest is recommended.

Daniel Byrd and Lester Lave consider de minimis risk policies as part of a broader risk management framework. The primary question they address is the coordination of a significant risk policy with cost-effectiveness considerations. As their title ("Significant Risk Is Not the Antonym of De Minimis Risk") suggests, their risk management framework includes a "nontrivial-but-significant" risk middle ground above de minimis and below significant for which costs and risk reduction opportunities are balanced. Byrd and Lave suggest that the significance of risks should be judged on the basis of the observability and uncertainty of health effects, using prior regulatory decisions as an aid to judgment.

Curtis Travis and Samantha Richter ("On Defining a De Minimis Risk Level for Carcinogens") and Paul Milvy ("De Minimis Risk and the Integration of Actual and Perceived Risks from Chemical Carcinogens") address the definition of de minimis risk from similar quantitative perspectives. Both chapters examine the risk levels associated with the variation in background levels of a number of carcinogenic substances; this approach is one of several that have been suggested for radiation. The central analysis in each chapter is the examination of the level of risk and population at risk for cases in which regulatory agencies did and did not regulate. For regulated substances, a further examination of the residual risk after regulation was conducted. Although Travis and Richter and Milvy present their data in somewhat different ways, both discussions found that when a lifetime risk above one per million is imposed on the entire U.S. population, a decision to regulate is usually made. They also found that the risk level associated with a decision to regulate increases as the population at risk decreases. For small populations at risk, lifetime risk levels of one in ten thousand are typically on the margin between regulation and nonregulation. The revealed pattern implies that regulatory decisions are sensitive both to the expected number of adverse health effects and to individual levels of risk. These chapters indicate a remarkable consistency in agency decisions when both factors are considered. These analyses are directly relevant to the definition of de minimis risk because they illustrate a fairly consistent pattern of response to small risks in recent regulatory decisions.

Gary Flamm, Robert Lake, Ronald Lorentzen, Alan Rulis, Patricia Schwartz, and Terry Troxell, all with the Food and Drug Administration, illustrate an inventive way of considering the potential severity of exposures to substances of unknown or uncertain potency. They start with the distribution of carcinogenic potencies observed in a large number of long-term animal toxicology studies and, with this distribution and a standard for lifetime risk, generate a distribution for permissible exposure. For low exposures, this analytical framework provides a basis for determining when testing is necessary, and when it is virtually certain that exposures to an untested substance are acceptably safe. Both the quantitative details and the logic of this analysis should have applications for low-level health risk protection in many contexts other than food safety.

The final chapters explore regulatory applications of the de minimis concept. Raymond Donnelly gives his view of how the de minimis concept has been successfully applied at OSHA, and how the OSHAct shapes the use of de minimis. A major aspect of OSHA's use of a policy equivalent to a de minimis policy is its use of programs that exempt firms from regular inspections, provided injury rates, inspection histories, and other factors indicate that exemption is warranted.

Sheldon Meyers provides an overview of how a de minimis philosophy can and has been applied to radiation at the EPA. He points out an important distinction in the EPA's use of a "below regulatory concern" (BRC) concept, in which regulatory concerns are established through a prioritization process that considers the finite resources of the agency and in which low risks are not regulated simply because to do so would use agency resources inefficiently. In contrast, de minimis traditionally refers to a value judgment that a risk is too small to be of social concern. Since it is often preferable to avoid defining what level of risk is acceptable or negligible, BRC is the preferred term in the EPA Radiation Program.

Charles Meinhold describes the factors that have gone into NCRP's committee study of de minimis. Of particular interest is the application of a negligible risk level that would serve as a lower limit to ALARA (As Low As Reasonably Achievable, a radiation protection term for cost-effectiveness). A negligible risk level is meant to be sufficiently small that individual and collective doses below this level would not be considered. A large number of comparative bases for judging the smallness of risk are discussed.

Miller Spangler and Joyce Davis provide detailed histories of the use of de minimis in radiation protection. Miller Spangler's chapter describes de minimis in relation to the regulatory policies of the Nuclear Regulatory Commission. Joyce Davis's study, conducted several years ago for the Edison Electric Institute, reviews radiation protection applications of de minimis from both a technical and a legal perspective, and describes how de minimis concepts have been used in response to a number of laws by several regulatory agencies. Both discussions include insightful critiques of many suggested approaches for establishing de minimis levels.

A number of thoughtful points were made at the workshop, some of which are not emphasized in these pages. The discussions at this workshop, as at most interesting meetings, served to illuminate the strengths and weaknesses of the presented papers. My interpretation of these discussions includes the following observations:

- No real controversy was raised regarding the definition of de minimis risk. Many terms were heard repeatedly—*trivial, insignificant, minimal*. However, each term is tailored for specific circumstances and reflects an overall framework for risk management.
- No one raised the concern that widespread implementation of a de minimis risk policy in regulation would lead to the creation or continued exposure to risks that otherwise would not be present.
- A major criticism of de minimis, provided by Richard Ayres, was that adoption of a de minimis risk policy would provide an undeserved moral legitimacy to the creation of small risks. This moral legitimacy would perhaps spill from the regulatory to the tort arena to limit compensation, and also could make later regulation more difficult. Proponents of a de minimis policy tended to emphasize its attractions in a regulatory context and did not address compensation issues, except to note that establishing causation for such small risks appears unlikely.
- There was general agreement that a universal de minimis risk number is not a practical concept. Important contextual factors, especially qualitative differences, should be considered. In this regard, Paul Milvy noted the significance to perceived risk of voluntariness, risk origin, severity of and time to impact, duration of risk exposure, familiarity, benefit or necessity, and distance. However, the chapters by Byrd and Lave, Travis and Richter, and Milvy indicate that there are some risk levels that have apparently never been regulated in any context.
- Much of the discussion concerned the activities of the regulatory agencies. The roles of the courts, Congress, and the Office of Management and Budget were not highlighted, although it seems quite likely that these institutions will have a major role in shaping how and when de minimis policies are established. Richard Guimond noted that the workshop focus was not *de minimis non curat lex* (the law does not concern itself with trifles) but instead was *de minimis non curat praetor*, referring to de minimis classification by the regulator rather than by the courts.
- The de minimis concept is potentially in conflict with cost–benefit approaches in several ways. This is apparent in the disagreement within the radiation community about whether de minimis risks are subject to or below ALARA, and in the tradition in de minimis proposals to focus on risk to individuals and in cost–benefit approaches to consider societal risk.

ACKNOWLEDGMENTS.   Many people helped to organize and conduct this workshop. The contributions of the workshop sponsors, identified at the beginning of the Preface, are very much appreciated. Janice Longstreth and Steve Swanson gave much time and effort on behalf of the National Capital Chapter of the Society for Risk Analysis; their efforts were critical to the successful organization and conduct of the meeting. Richard Guimond was both workshop cochair and an active program organizer. Both assistance and ideas were contributed by Richard Cothern, Vincent Covello, Billy Mills, Alan Moghissi, Melinda Renner, and Miller Spangler. Alvin Young, as chairman of the Committee on Interagency Radiation Research and Policy Coordination (CIRRPC), contributed in many ways, including service as a session chair and speaker about the activities of CIRRPC. I thank Peter Hutt, James DeLong, Richard Ayres, James Childress, Paul Price, and those speakers who contributed chapters to this volume. I thank Pauline Hostettler for her many contributions to the preparation of the proceedings.

Chris Whipple

Palo Alto, California

Introduction

Risk: A Pragmatic De Minimis Approach

Cyril L. Comar

Society is becoming increasingly well informed and anxiety-prone about technology-associated risks, which leads to desire for their elimination. The logical and traditional approach is first to estimate the risk, a scientific task. Then comes the issue of risk acceptance, a most difficult step—moving from the world of facts to the world of values. Ideally, judgments involving risk acceptance should be made on society's behalf by a constitutionally appropriate body. But no such public decision-making process exists. We make do with disparate efforts of individuals, special-interest groups, self-appointed public-interest groups, and legislative, judicial, and regulatory systems. However, if at least very large and very small risks were dealt with on the factual basis of effects, the individual and social value systems could be accommodated to some degree and much confusion avoided.

It is human nature to be concerned primarily with effects on our own person and family and secondarily with effects on the population at large. Unfortunately, although we can predict statistical effects on populations, there is no way to predict effects on individuals. This is why fortune-tellers never become as rich as insurance companies. We need then to define actuarially the existing state of well-being and calculate effects on it.

Each person has a probability of dying in any particular year, the value depending mainly on age. The existing probabilities are well known for the United States. For example, in 1975, 1.89 million died out of a population of 213 million, giving an overall probability of 1 in 113. For some specific age groups the values were these: 1 to 4 years,

The late *Cyril Comar* was Professor Emeritus, Cornell University, and Director, Environmental Assessment Department, Electric Power Research Institute, when this was originally published.

Reprinted with permission from *Science*, Vol. 203, Number 4378, p. 319, 26 January 1979. Copyright 1979 by the AAAS.

1 in 1425; 5 to 14 years, 1 in 2849; 25 to 34 years, 1 in 692; 55 to 64 years, 1 in 67. We can now answer the question, What does changing a risk do to a person's existing probability of dying? For instance, if a young child were exposed to an additional risk of 1 in 100,000 (0.014 in 1425) in 1975, his overall risk for that year would be 1 in 1425 plus 0.014 in 1425, or 1.014 in 1425.

For the purpose of discussion, some guidelines, which may depend somewhat on age, can now be stated in terms of numerical risk:

1. Eliminate any risk that carries no benefit or is easily avoided.
2. Eliminate any large risk (about 1 in 10,000 per year or greater) that does not carry clearly overriding benefits.
3. Ignore for the time being any small risk (about 1 in 100,000 per year or less) that does not fall into category 1.
4. Actively study risks falling between these limits, with the view that the risk of taking any proposed action should be weighed against the risk of not taking that action.

Clearly, these suggested guidelines are a gross oversimplification. The unfortunate, overtaken by a one-in-a-million catastrophe, have a 100% chance of harm. The hard fact is that attempts to eliminate risks for the unfortunate few tend to markedly increase them for the rest of a large population. This idea is most difficult to defend politically, especially when the unfortunate few are known and the unfortunate many are nameless. In addition, it is necessary to take into account such matters as validity and uncertainty in risk estimates, nonlethal and esthetic effects, voluntary versus involuntary risks, societal abhorrences, and the strange versus the familiar.

Nevertheless, other than depriving the news media of a ready source of attention-grabbing items, the pragmatic de minimis approach should serve to promote understanding about how to deal with risk in the real world; encourage identifiers of risk to provide risk estimates; focus attention on actions that can effectively improve health and welfare and at the same time avoid squandering resources in attempts to reduce small risks while leaving larger ones unattended; and prevent anxiety, apathy, or derision as a response to the increasing recognition that we are apparently live in a sea of carcinogens (the "today" risk).

Contents

I

De Minimis Risk Regulation
Incentives and Obstacles

1

De Minimis Risk
From Concept to Practice

Joseph Fiksel

INTRODUCTION

The statutes that govern federal regulation of environmental and technological risks represent the attempts of Congress to establish intelligible goals and principles as a guide for regulatory agencies. However, the statutory language frequently includes absolute statements that cannot realistically be satisfied or ambiguous statements that provide inadequate guidance. As a result, the regulatory process has often been hampered by difficulties in interpreting the intent and substance of the statutes, particularly in the context of low-level chronic risks.

One of the most challenging areas in statutory interpretation is the problem of setting cutoff levels for risk regulation. The consistency and effectiveness of risk management decision making might be enhanced if agencies had available a systematic approach for determining whether specific risks are de minimis—that is, too trivial to warrant the expenditure of resources for assessing or controlling them. However, development of a formal de minimis approach will require resolution of a number of important legal, technical, and policy issues that are merely touched upon here.

While de minimis risk has not been directly addressed by Congress, the de minimis concept is one that is well grounded in American common law. The Supreme Court and some other federal courts have recently exhorted agencies to consider the significance of risk findings in cases involving OSHA, EPA, and FDA. EPA has sought to establish *trigger levels* for regulatory action in areas such as hazardous waste cleanup or control of carcinogens, and has frequently designated specific risk levels as being "below reg-

---

Joseph Fiksel • Teknowledge, Palo Alto, California 94303. Dr. Fiksel was with Arthur D. Little, Inc., when this chapter was written.

ulatory concern." FDA appears to have invoked a de minimis rationale in recent actions, using a cutoff level of 10^{-6} individual lifetime risk. NRC is also considering a number of proposals for establishing a de minimis level of radiation dose that would be considered negligible. However, the de minimis approach remains a concept rather than a practice, and has not been adopted as a formal policy by the agencies.

There are a number of motivations for development of a rational and systematic approach to setting de minimis risk thresholds. These include the impossibility of completely eliminating risks without undesirable economic consequences, the multitude of public concerns in various risk sectors, the increasing sensitivity of chemical detection techniques, the profusion of suspected carcinogens and other chronic toxicants, and the desire for priority-setting mechanisms to assist agency decision making. The de minimis approach may represent an acceptable means of compromise between the desire to reduce risks and the desire for efficient utilization of resources.

THE DE MINIMIS RISK CONCEPT

The concept of de minimis risk has its origins in the legal principle *de minimis non curat lex;* i.e., the law does not concern itself with trifles. In the regulatory context, a risk is considered trifling or negligible if it is so trivial that the costs of regulatory consideration outweigh the importance of risk. More precisely, a risk is de minimis if the incremental risk produced by an activity is sufficiently small so that there is no incentive to modify the activity. A de minimis risk level would therefore represent a cutoff, or benchmark, below which a regulatory agency could simply ignore alleged problems or hazards.

The concept of de minimis risk is essentially a threshold concept, in that it postulates a threshold of concern below which we would be indifferent to changes in the level of risk. There are many analogous concepts in the natural and social sciences. For example, we can actually measure psychological thresholds of sensory perception, and we can experimentally demonstrate biological thresholds for toxic effects. In these instances, the threshold is an inherent property that can be discovered and quantified through empirical methods.

On the other hand, in dealing with societal issues in the economic, legal, or regulatory arenas, thresholds are not so easily identified; they depend to a large extent on judgmental interpretation and upon specific contextual circumstances. In particular, the selection of a de minimis risk level is contingent upon the nature of the risk, the stakeholders involved, and a host of other contextual variables. Thus, at the onset, it is clear that there is no universal or fundamental de minimis number. Rather, de minimis levels will be fuzzy, in that they can never be precisely specified, and relative, in that they will depend upon the circumstances.

The relativistic nature of de minimis risk is a crucial feature of this concept and can be demonstrated through a quasi-economic argument. Contrary to the beliefs of certain evangelical types, zero risk is neither achievable nor desirable. To rid society of most risks would require the elimination of many technological benefits and the suppression of innovation, leading to a decline in social welfare and a general stagnation. In such a scenario, the introduction of a small incremental risk might be welcomed if it were accompanied by significant technological benefits. Conversely, in a society subjected to multiple cumulative risks, the degree of anxiety about introduction of small incremental risks would be considerably higher. Thus, the level of concern assigned to a marginal change in risk depends on the overall existing risk level (actual or perceived), as well as upon the demand for benefits of risk-generating activities.

This does not imply, however, that a de minimis risk level involves a trade-off between risks and benefits. In fact, a key property of de minimis risk is that we would not be concerned about the introduction of a new activity with a risk below that level, no matter how slight the anticipated benefits. In contrast, the concept of acceptable risk involves a balancing of risks against benefits and a decision to accept a risk level that might not be tolerated if the benefits were smaller. In effect, de minimis risk is a lower bound on the range of acceptable risk for a given activity.

DE MINIMIS RISK IN PRACTICE

In a recent National Science Foundation study, several specific application areas were selected in order to illustrate the opportunities for use of a de minimis approach and the technical limitations that arise.[1] For example, the following observations were made:

OSHA Regulation of Benzene. One of the most significant features of the recent history of the benzene case is the role that OSHA has taken in the negotiation of a new exposure standard. The Supreme Court decision in *AFL-CIO v. API* reflected the view that where risks are highly uncertain, workers should not necessarily be assigned the benefit of the doubt. Although employers are generally required to bear the costs of reducing risks to levels established via regulation, the benzene decision suggests that workers may be expected to share the burden of uncertainty if the existence and magnitude of risks has not yet been determined.

The benzene case illustrates that it is difficult to set and enforce a de minimis risk level when carcinogens or suspected carcinogens are involved. Permissible concentration levels do not serve as a proxy for *risk* alone but instead combine a *risk* component with a *confidence level* component. By setting permissible levels sufficiently low, an agency can guarantee that the *probability* that the resulting risk level exceeds a given target level is small. This suggests that the de minimis concept needs to be extended to deal with uncertain risk, in order to be useful in regulatory decision making.

EPA Regulation of Benzene. It is instructive to see how the same hazard is dealt with in the context of benzene emissions discharged to air, under Section 112 of the Clean Air Act. On the basis of studies linking occupational exposure of benzene with leukemia, EPA's general presumption that carcinogenic risk thresholds do not exist, the absence of a demonstrated threshold for benzene, and widespread exposure to large quantities of benzene emitted by stationary sources, EPA proposed emission standards for four categories of sources; however, EPA withdrew three of the proposed standards because the health risks and potential risk reductions associated with those source categories were insignificant. This judgment was confirmed in the agency's final action withdrawing the proposals, which involve a de minimis-like argument.

On the other hand, EPA's determination that the residual risks remaining after a BAT standard is applied to fugitive emissions is *reasonable* in light of the costs of additional control involves an explicit trade-off rationale, and suggests that the agency has interpreted "an ample margin of safety" to mean something less stringent than a requirement that the probability of exceeding zero risk be very small. The implicit rationale for regulatory control in this case appears to be based on economic efficiency rather than on the trifling nature of the risk. This suggests that de minimis risk levels are not to be confused with less stringent standards based on the benefits per unit of the regulated activity, as well as on costs of risk control per unit of risk reduction.

CONCLUSIONS

The above examples illustrate that, while the notion of de minimis risk appears straightforward in principle, in practice it raises a number of analytical and policy issues. In order to develop a sound technical basis for setting numerical de minimis risk levels, the following fundamental principles should be considered:

- Designation of a risk as de minimis is meaningful only in the context of the activity that produces it and to the party that bears it.
- Establishment of de minimis cutoffs might be accomplished through surrogate measures, such as exposure or emission levels, which are more directly observable and quantifiable than risk itself.
- If commonly encountered per capita risks are to be used in selecting benchmarks for comparison, a distinction must be preserved between those risks that are assumed voluntarily, those that are accepted as part of a negotiated agreement, and those that are imposed upon third parties.
- A de minimis risk cutoff should not depend upon risk-benefit balancing considerations. Rather, de minimis risk should reflect a lower bound on acceptable risk levels, no matter what the associated benefits.

In addition, selection of de minimis risk levels that incorporate societal objectives such as equity, efficiency, or paternalistic intervention must consider a number of complicating issues:

- The need to balance individual per capita risks against cumulative risks for different numbers of exposed populations.
- Public aversion to catastrophic risks (e.g., accidental spills) as opposed to widely distributed risks (e.g., statistical cancer incidence).
- Availability of nonregulatory means (e.g., tort-law) for managing the residual risks that fall below the cutoff.
- Potential for unacceptable accumulation of residual risks owing to multiple de minimis exemptions.

While the above issues remain to be resolved, it appears feasible for federal agencies to incorporate de minimis risk criteria into their regulatory activities. On a practical level, those who invoke the de minimis concept must recognize that there is no universal magic number; the cutoff selected will depend upon the nature of the risk, the statutory context, and the policy objectives of the agency. Nevertheless, the systematic use of a de minimis approach can help to expedite agency decision making and to focus agency resources on nontrivial problems.

ACKNOWLEDGMENTS. The author gratefully acknowledges the substantive contributions of Michael Baram and Tony Cox, and the support of the National Science Foundation, particularly Joshua Menkes and Vincent Covello.

REFERENCE

1. Fiksel, J., M. S. Baram, L. A. Cox, and J. R. Miyares, "Principles for Use of De Minimis Concepts in Risk Regulation," Arthur D. Little, Inc. Final Report to Division of Policy Research and Analysis, National Science Foundation, Washington D.C., November 1984.

2

De Minimis Risk as a Regulatory Tool

Joshua Menkes and R. Scott Frey

Management of the health, safety, and environmental risks associated with various hazards has become increasingly difficult. To a large extent this difficulty can be traced to the fact that agencies are confronted with a seemingly unlimited number of risks and they have limited resources for managing such risks. One means proposed for dealing with this management problem is based on the legal principle *de minimis non curat lex* or, "the law does not concern itself with trifles;" hence the term "de minimis risk." According to this principle, risks considered trifling can be eliminated from regulatory consideration. Although the de minimis approach has been the subject of increased interest and agency use, the problems associated with its use need to be more fully addressed. In this chapter we (1) describe the concept of de minimis risk, (2) discuss several problems associated with the de minimis concept, and (3) conclude that the viability of the de minimis risk concept is not as compelling as some analysts have suggested.

DE MINIMIS RISK

The concept of de minimis risk is a pragmatic decision rule for distinguishing between trivial and nontrivial risks. This task is accomplished by establishing nonzero risk cutoff levels. If the risk of a hazard is greater than the de minimis level it becomes an object of further inquiry and possible regulation; however, if the risk falls below the de minimis level it is excluded from further consideration. It is argued by its proponents that such a decision rule relieves agencies from allocating scarce resources to trivial risks so that attention can be directed to important risks (Fiksel, Baram, Cox, and Miyares 1984; Whipple 1984).

Joshua Menkes • Lagos, Portugal.   *R. Scott Frey* • Department of Sociology, Anthropology, and Social Work, Kansas State University, Manhattan, Kansas 66506.

Proponents of the approach have suggested a variety of criteria that can be used to establish de minimis risk levels. Comar (1979), for instance, has proposed that de minimis levels be set at 10^{-5} annual level of mortality if the risk "carries no benefit or is easily avoided." Others have suggested similar numerical guidelines (cf. Fiksel et al. 1984). The bootstrapping approach has also been suggested as a means for setting de minimis levels. Specific criteria have included natural background levels in the case of radiation (cf. Whipple 1984) and risk levels of various hazards that are commonly encountered in daily life (cf. Fiksel et al. 1984; Mumpower 1984; Whipple 1984). In short, suggested techniques for establishing de minimis levels are based on either professional judgment or bootstrapping.

Discussions (e.g., Fiksel et al. 1984) of the de minimis approach indicate that it is consistent with current health and safety statutes and that it is well grounded in common law practice. In turn, agency efforts to establish insignificant risk levels in the assessment of suspected hazards are common. The fact that such levels have been labeled insignificant rather than de minimis is not important, for the logic underlying both concepts is the same. On the other hand, several agencies (including, for example, the Food and Drug Administration and the Nuclear Regulatory Commission) have adopted an explicit de minimis approach (cf. Fiksel et al. 1984; Spangler 1985; Whipple 1984). In sum, the de minimis concept not only has legal precedent but is currently being used by a number of regulatory agencies in the risk management process.

UNRESOLVED PROBLEMS

Despite its pragmatic appeal and its increased use, the de minimis approach is characterized by several unresolved problems (see, for instance, Fiksel et al. 1984; Mumpower 1984; Whipple 1984). Two problems of special interest here are (1) the viability of existing procedures for setting de minimis cutoff levels, and (2) the likely cumulative effects of the acceptance of many de minimis risks over time. We briefly examine these problems and conclude that they are not easily resolved.

Establishing De Minimis Cutoff Levels. Several proponents of the de minimis approach have suggested general numerical guidelines, such as the annual probability of mortality for exposed individuals, as a definition of de minimis risk (e.g., Comar 1979). As several analysts have noted (Mumpower 1984; Whipple 1984), this approach is problematic because it does not take into account the nature of the relationship between risk level and the size of the population exposed. In many cases, the correlation between risk level and exposed population is negative; i.e., small populations may be exposed to high-level risks or large populations may be exposed to low-level risks. If the correlation is negative, then the total number of fatalities resulting from a de minimis risk that affects a large population may be greater than those resulting from a risk that exceeds an identified de minimis level but affects a small population. In other words, a low-level or de minimis risk that affects a large population may be unacceptable because of the large number of fatalities associated with it.

For illustrative purposes consider the highly simplified constitutive relation between the number of individuals exposed to a given hazard: 10^a; the annualized probability of a fatality attributable to the hazard to an individual in the exposed population: 10^b; and the total number of fatalities expected per year: 10^c. The parameter b can be thought of as a "de minimis index" while c is a "catastrophe index." The constitutive relation is $10^a \times 10^b = 10^c$ or a + b = c. Specifying any two of the parameters a, b, or c determines the third. The nature of the relationship between the parameters is illustrated in Table 1.

Example I:
If the exposed population is 1000 then a = 3; if the fatalities per year are 10^{-3}, which corresponds to one fatality per thousand years, then c = -3. For these parameters:

$$3 + b = -3; b = -6.$$

Example II:
If we increase the exposed population to 10^4 but keep the expected number of fatalities the same then:

$$4 + b = -3; b = -7.$$

Example III:
If the exposed population is 250 million or $2.5 \times 10^8 = 10^{8.4}$ then a = 8.4. The expected number of fatalities is kept at 10^{-3} and this yields:

$$8.4 + b = -3; b = -11.4.$$

In all three examples the catastrophe index was kept constant at -3, which drove the de minimis index from -6 in the first example to -11.4 in the third example. The point is that the size of the exposed population is an important component in the establishment of de minimis cutoff levels. In effect, small individual probabilities are translated into large numbers of fatalities because of the large number of individuals at risk. This consideration applies to discrete risks as well as to chronic risks.

The *size* of the exposed population is an important consideration in the establishment of de minimis cutoff levels as well as the *number* of hazards that would be left unregulated as a result of the widespread application of the de minimis approach. It is quite easy to

Table 1. Constitutive Relation between
Exposed Population, Probability of Fatality,
and Fatalities per Year[a]

a	b	c
2	-5	-3
3	-6	-3
4	-7	-3
8	-6	2
8.4 (Total U.S. population)	-11.4	-3
$10^a \times 10^b = 10^c$		

[a] a = exposed population, b = probability of fatality (de minimis setting), c = fatalities per year.

imagine a situation in which a large number of hazards are trivial when considered individually, but when such hazards are considered collectively they may be quite risky to exposed individuals. Small individual probabilities are translated into large numbers of fatalities if the de minimis logic is applied to a large number of hazards. The point is that the total number of risks covered by application of the de minimis approach is as important in setting de minimis risk levels as the number of exposed individuals.

Even if such problems could be adequately resolved, an additional problem remains: Who has the authority to define specific numerical de minimis guidelines that can be applied to a wide variety of risks? In other words, good reasons have to be given for the decision to select one numerical guideline over another. Many of the existing specifications fail to provide such justifications, for they appear to be based solely on professional judgment. Consequently, efforts to define numerical guidelines such as 10^{-5} annual level of mortality as de minimis are suspect.

An alternative to the practice of establishing general numerical guidelines applicable to a wide variety of risks is the use of the bootstrapping method. According to this view, de minimis risk is defined in terms of risks that have been tolerated in the past. That is, de minimis levels for specific risks are based on the consideration of risks that are commonly accepted as a part of everyday life. For instance, natural background levels of radiation have been suggested as a means for establishing de minimis standards for human exposure to radiation. Other criteria such as revealed and implied preferences have also been suggested as means for establishing de minimis cutoff levels for specific risks.

Although the bootstrapping method is more defensible than the alternative of professionally based guidelines, the practice is not without its problems. One of the most obvious problems with this method is its reliance on current patterns of risk. Such a position assumes the legitimacy of the political, economic, and social relations that underlie extant patterns of risk. In other words, this particular technique is based on the questionable assumption that what is, ought to be.

It is clear from this discussion that the practice of establishing de minimis cutoff levels is plagued by several problems. These problems are so serious that they may prove to be intractable. The question of intractability aside, the point remains that the de minimis approach will remain a concept rather than a widely implemented practice until the problems associated with setting de minimis cutoff levels have been adequately resolved.

Cumulative Effects of Multiple Risks. A risk management strategy that allows many de minimis risks to accumulate over time may increase overall risks more than a strategy that allows only a few nontrivial risks to accumulate. This is the case because multiple low-level risks can interact with one another in various ways to produce risks very different from those produced by any single one.

The problem is magnified when we consider the nature of the risk management process itself. It is a process characterized by a large number of regulatory agencies. Each of these agencies is constrained by jurisdictional limits on the range of its functional concerns and responsibilities. Functions can be defined in such a way that an agency is not allowed to consider risks occurring outside of its domain. In effect, there is a tendency to view these risks as someone else's problem. The result is that the acceptance of many de minimis risks over time by different agencies following their mandated responsibilities may lead to an increase in overall risk.

Despite the problems regarding the viability of setting de minimis levels, the most important problem with this approach is its possible consequences. Far from limiting risks in an efficient fashion, implementation of the de minimis approach might actually increase risks. This is an unresolved difficulty that weights heavily against the wide-scale implementation of the approach.

CONCLUSIONS

As we noted at the beginning of this chapter, the management of the health, safety, and environmental risks associated with various hazards has become increasingly difficult because regulatory agencies have limited resources and they are confronted with a seemingly unlimited number of risks. Given such a situation, the de minimis approach would appear to be an ideal decision tool for guiding regulatory actions in an efficient and judicious fashion. However, the de minimis approach is characterized by two serious problems: the viability of existing procedures for setting de minimis cutoff levels, and the cumulative effects of the acceptance of many de minimis risks over time.

Each of these problems is sufficiently serious to suggest a critical evaluation of the applicability of the de minimis approach to risk management. In fact, wide-scale implementation of the approach may have undesirable consequences that are long-term and irreversible. Additional work addressing problems raised in this chapter is needed before the de minimis approach can be considered a plausible strategy for effective and efficient risk management.

REFERENCES

Comar, C. L., "Risk: A pragmatic de minimis approach," *Science* 203: 319, January 26, 1979.

Fiksel, J., M.S. Baram, L.A. Cox, and J.R. Miyares, *Principles for Use of De Minimis Concepts in Risk Regulation,* Arthur D. Little, Inc., Reference 50363, Final Report to Division of Policy Research and Analysis, National Science Foundation, 1984.

Mumpower, J., "Analysis of the de minimis strategy for risk management," Paper presented at the annual meeting of the Society for Risk Analysis, Knoxville, 1984.

Spangler, M., "A summary perspective on NRC's implicit and explicit use of de minimis risk concepts in regulating for radiological protection in the nuclear fuel cycle," Paper presented at the De Minimis Risk Workshop, Brookings Institution, Washington, D.C., May 29–30, 1985.

Whipple, C., "Application of the de minimis concept in risk management," Paper presented at the Joint Session of the American Nuclear Society and Health Physics Society, New Orleans, June 6, 1984.

3

Application of the De Minimis Concept in Risk Management

Chris Whipple

INTRODUCTION

In risk regulation and management, de minimis refers to the general theme that some risks are too small to be of societal concern. It is easy to show that all organizations with risk management responsibilities use some type of de minimis approach, since risk management resources are always finite, and the supply of very small risks virtually inexhaustible. To cite just one example, the prohibition of carcinogenic food additives under both the Delaney Clause and earlier food safety laws is commonly cited as a "zero risk" policy statement that is as protective as any risk policy on the federal level. Even so, no serious consideration has ever been given to the idea that the Food and Drug Administration should ban food additives that contain a radioactive molecule or two, since that would include virtually all substances. While all organizations with risk management responsibility follow some pragmatic de minimis approach, there has been a series of proposals (see Chapter 13 and Davis[1] for a general review of the issue) for explicit adoption of the de minimis concept by regulatory agencies.

The purpose of this chapter is to consider two issues relating to the use of de minimis levels as cutoffs for regulatory concern. First are the foreseeable advantages and drawbacks of establishing an explicit de minimis policy in contrast to *ad hoc* or other pragmatic risk management approaches. Second, several practical and philosophical problems associated with alternative de minimis approaches are considered. In both cases, the de minimis

This paper was originally presented to a Joint Session of the American Nuclear Society and the Health Physics Society, New Orleans, June 6, 1984.

Chris Whipple • Energy Study Center, Electric Power Research Institute, Palo Alto, California 94303.

15

approach is considered in a general sense, that is, as a tool applicable to all types of risk, not just radiation.

There are several reasons for considering the de minimis approach in a general context suitable for all low-level risks. First, the considerations that apply to de minimis in a radiation context typically apply in other situations. In many cases the degree of risk is known to be low, but large uncertainties preclude precise estimates. Often, large populations are exposed. ften, public or political interest in the issue arises owing to common qualitative features, e.g., involuntary exposure to a carcinogenic agent. The similarity of these characteristics, found with food safety and environmental issues as well as with radiation protection, suggests that workable solutions may have common features. Second, it can be efficient to adopt a common approach to different risks. Given limited resources, it makes sense to pay attention to large risks and to risks that are easily controlled. A common definition of de minimis can reduce the variation in regulatory response to comparable risks. Finally, utilization of formal de minimis rule for many risks may enhance the acceptability of the approach.

INCENTIVES TO USE A DE MINIMIS APPROACH

The impetus to establish a consistent de minimis approach to risk regulation has increased in recent years for several reasons. First, technologies for identifying risks have improved in several ways. Improvements in analytical chemistry permit the detection of hazardous substances at the part-per-billion or even part-per-trillion level; only a decade ago such exposures would have been ignored simply because they would have been undetectable. (Radiation is an exception to this rule, since it has been detectable at low levels for decades. This helps to explain why so many de minimis proposals arose from the radiation protection area.)

Second, the likelihood that any specific chemical would be suspected of posing a risk has increased with the advent of *in vitro* and other fast and inexpensive tests. Our view of the nature of low-level risks has shifted somewhat over the past decade. The view that we face rare but potent carcinogens seems to have given way to the view that carcinogens (at least as determined by high-dose animal tests and by inference from *in vitro* bioassays) are fairly commonplace, and significantly varied in their carcinogenic potential. Recent work that reveals widespread exposures to natural carcinogens in food[2] has troubling implications for those who favor the elimination of carcinogens as a risk management policy (see reference 3) and strengthens the argument to move to a carcinogen management approach that prioritizes regulatory attention based on both carcinogenic potency and exposure.

An impact of the increasing number of candidate substances for regulation and of the apparent need to prioritize regulatory efforts is that case-by-case decision making is seen as too cumbersome. As the regulatory approaches to low-level agents mature, more systematic methods are sought. To regulators, the de minimis approach appears to provide a means of normalizing the process, by providing an alternative to standard setting for substances that pose very low risks. This role is particularly important where an agency has a statutory mandate to deal with a substance or class of substances, but where resources are not available to deal with low-risk and low-priority substances.

In addition to these regulatory incentives for using a de minimis approach, industry is likely to be supportive of a de minimis approach since it defines a threshold for regulatory involvement. The de minimis rule could produce greater predictability and stability in regulation and could provide a check in the rare cases where a relatively safe substance meets a regulatory approach disproportionate to the risk (e.g., cyclamates[4]).

To the scientific community, the de minimis approach may provide a policy solution to questions that lie beyond the reach of scientific resolution. This would reduce the pressures for regulatory agencies and their scientific staffs to produce scientific judgments regarding low-level risks where information is unavailable.

An additional incentive for a de minimis approach comes from recent court decisions indicating that a regulatory agency should determine that a risk is significant before setting regulations. The Supreme Court was split on this issue in the OSHA benzene case, but the point was rendered moot owing to other considerations (see reference 5). The D.C. circuit court supported the de minimis position in the case of trace quantities of packaging materials migrating into food (the case involved the use of acrilonitrile). More recently, the EPA announced plans to permit use of the pesticide Larvadex under a de minimis approach previously proposed by FDA.[6] (The Larvadex policy was revised when new risk data became available.)

SOCIAL ACCEPTANCE OF DE MINIMIS RISK MANAGEMENT

A major consideration with the application of a de minimis policy to risk management is the effect such a policy would have on risks borne by the public or an occupational group, in comparison to the risks produced in the absence of a de minimis policy. Clearly, many small risks that would be formally excluded from regulatory concern under a de minimis approach are unlikely to be regulated under any approach. These risks are not the issue, since in such cases the de minimis policy would make no difference.

A politically contentious issue arises when a de minimis policy is considered for risks regulated under a statute that appears to require regulatory action for a class of risks, regardless of the degree of risk. The commonly cited examples of such laws include the Food and Drug Laws, especially the Delaney Amendment, which prohibits food additives that "induce cancer when ingested by man or animal," and sections of the Clean Air Act that require standards to protect public health (including that of sensitive subgroups in the population) with an adequate or ample margin of safety.

While the residual risks permitted under a de minimis rule would be quantitatively small by definition, the public reaction to such risks may be influenced more strongly by the qualitative characteristics of the risk.[7] For many of the agents for which the de minimis approach is being considered, the risk is uncertain and carcinogenic; these characteristics are among those commonly cited as enhancing the degree of public concern to a risk.

It is not easy to predict the public reaction to a de minimis system only on the basis of an abstract concept, since the reaction is more likely to depend on initial reactions to a few well-publicized incidents. One can envision how a de minimis approach could encounter substantial public opposition if there were identifiable victims of some de minimis exposure. Just as the Ford Pinto gas tank design decision has made automotive engineers cautious about using cost-effectiveness analysis, a well-publicized "bad" outcome under a de minimis rule could restrict this approach. Conversely, de minimis could be popularly received if it was seen as a means of pragmatically avoiding some rigid regulatory action perceived to offer little public benefit (e.g., OSHA requirements for the height of toilet seats).

CONFLICTING SOCIAL OBJECTIVES

Our objectives in risk management involve fundamental conflicts that rule out a completely satisfactory approach.[8] One widely espoused value statement is the simple desire to have safety—that is, to eliminate risks to the extent that is possible. Countering this social objective is the desire for efficiency in risk management. This desire for efficiency rests on the argument that resources are scarce, but as Calabresi and Bobbitt have noted,[8] "commonly . . . , scarcity is not the result of any absolute lack of a resource but rather of the decision by society that it is not prepared to forgo other goods and benefits in a number sufficient to remove the scarcity."

In health and safety risk regulation, the conflicting objectives of near absolute protection and careful use of scarce resources have produced approaches in which these two objectives are considered in varying degrees. In many cases, economic efficiency is explicitly stated as one of several regulatory goals, and the analytical examination of regulatory costs and benefits is customary or even obligatory. In such cases, the de minimis approach is likely to formalize the practice of ignoring very small risks. For other risks, the regulatory mandate is more strongly focused on protection; in these cases the costs of achieving safety are secondary considerations or not legal considerations at all, at least in principle. In practice, cost considerations do usually influence all decisions to some degree. It seems possible that de minimis is particularly useful, because of its historical legal acceptance,[1] to avoid the regulation of trivial risks that a literal reading of the law would require. Under a well-designed de minimis system, the social objective to have a high degree of safety could be met, while at the same time the need to ignore small risks could be recognized. In this role, and under favorable circumstances, de minimis offers the possibility of bringing practical considerations into decisions where very small risks are involved without resort to the risk–cost trade-offs that many find offensive, and that certain laws prohibit. In short, de minimis may, in certain cases, permit us to avoid facing the difficult conflicting objectives contained within our risk value system.

On a larger scale, the practical significance of a de minimis policy may result from the pressures it creates to reevaluate inconsistencies in the attention applied to various risks. A de minimis risk policy is philosophically consistent with the view expressed by Lord Rothschild:[9] "There is no point in getting into a panic about the risks of life until you have compared the risks which worry you with those that don't but perhaps should."

A second basic objective in modern risk regulation is the separation (to the extent that is practical) of questions of science from questions of policy.[10] This objective arises from several motivations, notably to permit public participation in the formulation of risk policy without requiring scientific expertise, and to permit scientific debate over the degree of risk to be conducted at a distance from questions of regulatory action. In practice, such separation is difficult to maintain, as illustrated by the tendency to use conservative assumptions in risk assessment when uncertainties are large. Such conservatism reflects a societal consensus to err on the side of safety in risk matters. This issue of separation is particularly difficult for very low risks, because uncertainties are so great at low levels. For many exposures there is no direct evidence that risk exists; the evidence is simply that risk exists for much higher doses. Out of prudence we assume that the risks do exist, that thresholds do not exist. But this raises further difficulties, as John Gibbons notes:[11]

> A zero-threshold situation leaves the policymaker in a great quandary. As long as there is some threshold level below which there are no ill effects, social equity can be preserved. But if dose and effect have a zero-zero intercept, then the policymaker must talk about determining acceptable risk, which is far more difficult to deal with than no risk.

In other words, risk management has become much harder because we no longer believe in thresholds, at least scientifically. Here the de minimis approach can offer policy thresholds in lieu of scientific thresholds.

DE MINIMIS VERSUS ALARA

The central de minimis concept, that some risks are too small to be of collective concern, is conceptually in conflict with the idea that no risk should be taken without some benefit, and that reasonable (i.e., cost-effective) opportunities for risk reductions should be sought for all risks. These latter considerations have been the central principles for radiation risk management recommendations of the International Commission on Radiological Protection,[12,13] and consequently the ICRP has not endorsed a de minimis concept.

While it appears attractive to consider a restricted definition for de minimis, where this term would refer to risks that are individually low, associated with some social benefit, and not easily eliminated or reduced, such a redefinition involves the complex considerations of risk, benefit, and risk control that the de minimis concept is meant to avoid.

A proposal for using a de minimis concept that is essentially consistent with ICRP principles has been proposed by researchers at the UK National Radiological Protection Board (NRPB);[14] the proposal is relatively restrictive given the constraints imposed by ICRP principles. Unlike many de minimis proposals based on the level of individual risk without regard to the size of the population group exposed, the proposal from NRPB considers both societal and individual risk. This proposal begins with the premise that individual risks producing an annual probability of death of 10^{-6} are typically not decision factors, so an additional risk of this magnitude to an individual from a collection of radiation exposures would not be excessive. An individual annual dose limit of 10^{-6} Sv (1 Sv = 100 rem) was proposed; this corresponds roughly to a risk of 10^{-8} per year. The additional factor of 100 was included to account for the possibility of exposures from multiple sources.

An additional constraint was proposed that such exposures be considered de minimis only if the collective dose to the exposed population was of the order of one man-Sv. This value was selected on the grounds that population doses below this level did not justify the analytical resources required for their consideration. Given a typical cost of £1000–10,000 for even a simple analysis, and an NRPB ALARA criterion of £2000 per man-SV avoided (about $30 per person-rem), it makes sense to assume *a priori* that it is a poor use of resources to consider population doses below this one man-SV level. As the authors note,[14] "in this context, the concept of de minimis reduces to nothing more than a specific form of ALARA appropriate for trivial radiological problems."

In comparison to simpler and less restrictive de minimis approaches based solely on ignoring individual risks that are below a specified level, the NRPB proposal includes the additional consideration that the sum of a very large number of very small risks can be significant. This consideration applies both to the case in which a single individual is exposed to a number of risks and to the case in which a large population is exposed.

INDIVIDUAL VERSUS SOCIETAL RISK

Certainly a society can manage risk to the population as a whole by limiting individual risks. This is, in fact, the approach taken by the Nuclear Regulatory Commission in its proposed safety goals for nuclear power plants.[15] Individual risk limits are appropriate in cases where individuals would face relatively high risks. But when individual risks are not inequitably high and the motivation for risk management comes more from having a large number of people facing a low to moderate risk than from having a few people at high risk, then individual risk approaches can produce resource misallocations. To take a hypothetical example, a 10^{-6} per year risk to 1000 people produces 10^{-3} expected fatalities per year, equivalent to an expectation of 1 fatality per 1000 years. This same risk of 10^{-6} per year applied to the entire U.S. population of 230 million produces an expectation of 230 fatalities per year. If a judgment of whether these were de minimis risks hinged on the individual risk, the regulatory response (or nonresponse) is likely to treat these risks comparably. Yet common sense tells us that it is almost certainly a poor use of safety resources to address the first risk but not necessarily the second.

The de minimis concept is precisely concerned with the definition of significance; therefore, it seems appropriate to include both individual and societal considerations in proposed definitions. Following the NRPB logic (see reference 14) (but not their numbers), it might be reasonable to define a de minimis risk as one that is less than 10^{-5} per year (probability of death) for a population of less than 10^3, less than 10^{-6} per year for exposed populations of 10^3 to 10^6, and 10^{-7} per year for the entire population. This suggestion places emphasis on individual risk levels for small populations at risk, and emphasis on expectation for lower individual risks to larger populations.

DE MINIMIS PROBABILITY

An additional question in defining a de minimis approach is whether one can define a de minimis probability in isolation from consideration of consequence. This issue has relevance to engineering risk calculations of several types. For example, the risks from airplanes flying over high population areas is generally ignored (except for cases where air traffic density is at issue), as are the risks to nuclear plants and other facilities from meteorites. The issue of a de minimis probability also bears on the selection of design criteria based on seismic risks. The distinctive feature of risks to which a de minimis probability might apply is that, unlike toxic agents in air, food, or water, such risks are considered catastrophic in that they cause simultaneous fatalities.

One analysis[16] of the irreducible uncertainties in risk indicated that estimates of less than roughly 10^{-10} per year for such risks are not meaningful, since "new" events (i.e., those not yet seen or postulated) set a lower bound at around this level.

More to the point, if one follows the NRPB approach of considering as de minimis those risks with expectation costs of the same order as the costs of analysis (perhaps of the order of $10,000–100,000), and further assuming that it is appropriate to represent a catastrophic risk by its expectation, then an accident with consequences of the order of $10 billion and probability of 10^{-6} per year would be de minimis. This $10 billion figure was chosen because it is higher than the estimated cost of damages associated with the worst historical accidents other than natural hazards.[17]

It is possible to find examples of risks around or even below these levels (e.g., severe nuclear power or commercial aviation accidents) that receive considerable analytical and managerial attention. It is also possible to find risks of similar scale or larger (e.g., earthquakes in Eastern U.S. cities, chemical plant explosions) that receive substantially less regulatory attention. (Although earthquakes are natural hazards, their risk is manageable through building codes.) As these cases suggest, it appears that generalizations about the willingness of the public to neglect potentially high-consequence, low-probability risks are difficult to draw. From the analytic points raised here, it appears possible to adopt a de minimis rule for this category of risk.

CHOOSING A DE MINIMIS LEVEL

A number of rationales have been suggested for the quantitative definition of a de minimis risk. (For radiation protection, alternative approaches are described by Davis.[1]) In some instances, these levels seem to be clearly labeled value judgments; e.g., Comar suggested a value of 10^{-5} per year,[18] without any explicit or comparative rationale.

Certainly no one disputes that defining a de minimis risk level is a social value judgment, and, as a consequence, there is no scientific definition of de minimis. But there are a number of logical arguments for a de minimis level, usually based on risk comparisons, that make use of quantitative risk estimates. Davis[1] reviews a number of proposed bases for selecting a de minimis level for radiation risk. Several proposed approaches, such as comparisons with background levels, are specific to radiation and are not easily generalized.

Some de minimis proposals[1] have a scientific rationale, which suggests that exposures below some level have little health significance. For example, virtually no risk would occur where there is a threshold dose below which no health effects are likely to occur, or where there is a latency period that decreases with dose and which, for a given dose, would be associated with a latency period that extends beyond the natural life span. Similarly, it has been argued that exposures to low doses of ubiquitous risk agents (e.g., radiation, nitrosamines, formaldehyde) are probably less harmful than exposures to un-common agents, owing to the likelihood of human defense or repair mechanisms against common risks. On this basis, comparisons with background may be appropriate.

Another proposed approach[1] is to set de minimis levels on the basis of consideration of the observability of health effects. Anticipated effects would be characterized as de minimis if they were not detectable epidemiologically. The obvious difficulty here is that the degree of risk associated with the de minimis levels varies substantially with the effect. A substance that causes a rare effect (e.g., vinyl chloride) would have a very low de minimis level, while one that produced a more common effect (e.g., lung cancer) could result in a de minimis risk level that would be considered significant in other contexts.

All of these proposed methods seem to have disadvantages that would preclude their adoption as general approaches to the definition of de minimis levels. The comparison of exposures with natural background for radiation provides reassurance that health effects will occur rarely, if at all. But this cannot be applied to many of the chemicals that are currently of concern because there is often little or no background exposure. The selection of a de minimis exposure level based on estimated thresholds or latency periods asks for more scientific evidence than can typically be provided. Further, the fact that the as-sumptions of threshold or long latency may be nonconservative when one considers multiple and cumulative exposures causes these approaches to appear prone to errors in which a greater number of health effects than anticipated could occur. Such an approach goes against the conservative policy generally applied to toxic materials.

A more commonplace reason for comparing risks is to provide insights to the social significance that is accorded to risks of various magnitudes. As applied to the de minimis issue this revealed preference approach seeks to define de minimis risks as those smaller than risks that are typically ignored. One such example was the proposal[19] to compare risk (or radiation exposure) with the variations in background radiation. The rationale here is that people could, but do not, consider these variations in decisions.

A logical requirement for using a risk comparison as a de minimis reference point is that avoidable risks be considered, because unavoidable risks may not be accepted so much as they are tolerated. However, as one considers candidate reference risks, it quickly becomes apparent that avoidability is virtually always a question of degree. An additional difficulty is that many avoidable risks are undertaken voluntarily, and consequently are categorically different from the types of involuntary risk[12] that are often at issue when de minimis is considered. The use of background radiation variation avoids both these objections.

A further complication in setting a de minimis level based on such comparisons arises from the observable inconsistencies in the degree of concern associated with risks of different levels. Clearly, the public concern with specific risks does not track actuarial risk.[7] So at any reasonable suggested de minimis level, lower risks that are of public concern can be found, and higher risks that are routinely ignored by a majority of the public can be cited.

If one were motivated primarily by public concerns regarding risk, an approach to de minimis might be developed that was flexible and sensitive to the nature as well as the degree of risk. Conversely, an approach to de minimis that emphasizes efficiency will focus on expectation to a greater extent than is now observed in de minimis proposals.

Current approaches tend to be based on individual risk probabilities,[1] excluding both population-at-risk measures and qualitative risk factors.

MULTIPLE SOURCES OF RISK

It has been noted[1,14] that the sum of a large number of small risks could be significant. It is easy to imagine how this could present a problem for the application of a de minimis concept by a number of narrowly focused regulatory bodies. A suggested solution to this problem[1,14] is to limit risks from a particular agent in a particular context to a fraction (e.g., 1%) of the de minimis limit. Davis[1] goes further and suggests that a monitoring program and analysis can indicate whether or not such multiple exposures defeat the purpose of the de minimis concept. In both these cases, the authors have protection against radiation in mind and are basing their recommendation on the objective of limiting the risk from a collection of exposures to radiation from a variety of sources. In the broader case where de minimis is considered as a policy for low-level risks generally, this problem of risk accumulation is conceptually similar, but operationally more difficult.

It is apparent that the definition of a de minimis risk should include consideration of the possibility that multiple de minimis exposures could result in a large aggregate risk. Assuming that a de minimis level is an individual risk limit from a particular agent and circumstance, then there are several contingencies that could arise. First, as noted above for radiation, there can be multiple sources of exposure to the same agent. For instance, one could conceivably be exposed to the same hazardous material in drinking water, by inhalation, or in a variety of foods. These are real possibilities if we consider as an example exposure to a pesticide in a rural area. As a practical matter, one or two pathways are likely to dominate exposures, and it seems unlikely that total exposures could be greater than several times the exposure via the most significant pathway. Given the uncertainties in risk estimates at such low levels, this factor seems fairly trivial.

A far more troubling question is posed by the sheer number of risk agents. It would hardly be comforting to learn that although no single chemical in drinking water poses a cancer risk in excess of 10^{-7} per year, there were 10^6 such chemicals. The problem this issue poses for the use of de minimis as a regulatory threshold is the degree to which risks are examined and managed singly versus on an aggregated basis.

While the prevalent approach of doing risk analyses on a chemical-specific basis seems to support a de minimis concept on an agent-by-agent basis, other arguments support a de minimis definition for an aggregation of agents. Such an approach provides confidence that the sum of de minimis risks any individual sees is limited by the level of aggregation. Clearly, there are many ways to aggregate: All effluents from a single facility could be considered, or one could consider the risk from all chemicals found in drinking water, for example. Such aggregation might provide an incentive to design effluent control systems on a basis in which risk-based trade-offs between multiple effluents are considered, e.g., an extension of EPA's bubble concept. Potential advantages of permitting such trade-offs between risks have been noted.[20]

As these cases are meant to illustrate, the approach taken under a de minimis philosophy to avoid excessive accumulations of risk can take many forms, depending on the specific context. Apparently the Food and Drug Administration has proposed a definition of a de minimis risk that does not treat this issue directly except through the use of a very low de minimis definition of 10^{-6} lifetime risk of concern.[21] The NRPB proposal described above arrived at virtually the same risk level, but in this case, a higher initial risk target (10^{-6} per year rather than 10^{-6} per lifetime) was modified by a factor (of 100) to account for accumulations.

SUGGESTIONS

Clearly, many issues must be resolved to develop a workable de minimis policy. Proposed de minimis approaches have dealt with some aspects of the problem, notably with the comparative logic that justifies selection of a de minimis level. More effort is needed, however, if the de minimis idea is to be adopted as a generally sound risk management approach. In particular, the aggregation of multiple risks has not received sufficient attention. Two additional ideas that deserve attention are the consideration of population risks where appropriate instead of individual risks, and the idea of de minimis probability.

Finally, it should be recognized that public acceptance of de minimis may be the critical constraint, and should be a consideration in any proposed approach. To that end, it may be helpful to think of de minimis risks as those that are of too low a priority to regulate, rather than as acceptably low risks. This distinction, by framing the issue as prioritization rather than acceptability, naturally encourages a comparative risk viewpoint and avoids the difficult question *acceptable to whom?* The proposal to define as de minimis risks those that cannot be measured epidemiologically would avoid creating identifiable victims of a de minimis approach. This approach might be considered a necessary condition for defining a de minimis level; however, it is not sufficient for the reason noted above.

An additional factor that would appear to promote the acceptability of a de minimis approach is the emphasis that, in general, de minimis levels apply not to actual known risks but rather to risks of unknown magnitude that are conservatively estimated. Ideally, one would like to be flexible in response to the likely degree of conservatism in a risk estimate.

REFERENCES

1. J. P. Davis, *The Feasibility of Establishing a De Minimis Level of Radiation Dose and a Regulatory Cut-off Policy for Nuclear Regulation*, General Physics Corporation Report GP-R-33040, Columbia, Maryland, December 31, 1981.
2. B. Ames, "Dietary Carcinogens and Anticarcinogens," *Science 221*(4617): 1256–1264, September 23, 1983.
3. S. S. Epstein, J. B. Swartz, et al., letter to *Science, V. 224*, N. 4650, May 18, 1984, and reply by B. Ames.
4. W. R. Havender, "The Science and Politics of Cyclamates," *The Public Interest, N. 71*, Spring 1983.
5. A. Scala, "A Note on the Benzene Case," *Regulation*, July/August 1980.
6. E. Marshall, "EPA Regulators Take on the Delaney Clause," *Science V. 224*, N. 4651, May 25, 1984.
7. B. Fischhoff, P. Slovic, S. Lichtenstein, S. Read, and B. Combs, "How Safe is Safe Enough? A Psychometric Survey of Attitudes Towards Technological Risks and Benefits," *Policy Sciences V. 8*, pp. 127–152, 1978.

8. G. Calabresi and P. Bobbitt, *Tragic Choices*, W. W. Norton and Co., New York, 1978.

9. N. Rothschild, "Coming to Grips with Risk," *The Wall Street Journal*, May 13, 1979.

10. Committee on the Institutional Means for Assessment of Risks to Public Health, National Research Council, *Risk Assessment in the Federal Government: Managing the Process*, National Academy Press, Washington, D.C., 1983.

11. J. H. Gibbons, in *Public Policy, Science, and Environmental Risk*, proceedings of a workshop at the Brookings Institution on February 28, 1983, edited by S. Panem, The Brookings Institution, Washington, D.C., 1983.

12. National Radiological Protection Board, *The Application of Cost-Benefit Analysis to the Radiological Protection of the Public: A Consultive Document*, Harwell, Didcot, Oxon. OX11ORQ, HMSO, March 1980.

13. Recommendations of the International Commission on Radiological Protection (adopted January 17, 1977). ICRP Publication 26, Ann. ICRP, 1, No. 3, Pergamon Press, Oxford, 1977.

14. R. H. Clarke and A. B. Fleishman, "The Establishment of De Minimis Levels of Radioactive Wastes," presented at the IRPA Congress, Berlin, 1984.

15. U.S. Nuclear Regulatory Commission, Office of Policy Evaluation, *Safety Goals for Nuclear Power Plant Operation*, NUREG-0880, Revision 1 for Comment, Washington, D.C., May 1983.

16. C. Starr, R. Rudman, and C. Whipple, "Philosophical Basis for Risk Analysis," in *Annual Review of Energy V. 1*, Annual Reviews, Inc., Palo Alto, 1976.

17. K. A. Solomon, "How Unique are the Price-Anderson Limitations on Nuclear Accident Liability?," *Risk Analysis, V. 3*, No. 1, March 1983.

18. C. Comar, "Risk: A Pragmatic De Minimis Approach," *Science, V. 203*, No. 4378 January 26, 1979, reprinted in this volume.

19. H. L. Adler and A. M. Weinberg, "An Approach to Setting Radiation Standards," *Health Physics, V. 34*, pp. 710–720, June 1978.

20. P. Huber, "The Market for Risk," *Regulation*, March/April 1984.

21. D. A. Kessler, "Food Safety: Revising the Statute," *Science, V. 223*, March 9, 1974.

4

Science and Its Limits
The Regulator's Dilemma

Alvin M. Weinberg

William Ruckelshaus, in his beautiful essay "Risk, Science and Democracy,"[1] has expressed very clearly what I shall call the regulator's dilemma. "During the past 15 years there has been a shift in public emphasis from visible and demonstrable problems, such as smog from automobiles and raw sewage, to potential and largely invisible problems, such as the effects of low concentrations of toxic pollutants on human health. This shift is notable for two reasons. First, it has changed the way in which science is applied to practical questions of public health protection and environmental regulation. Second, it has raised difficult questions as to how to manage chronic risks within the context of free and democratic institutions."[1]

When the concerns were patent and obvious—like smog in Los Angeles—science could and did give unequivocal answers: For example, smog comes from liquid hydrocarbons and the answer to smog lay in controlling emissions of these substances. The regulator's course was rather straightforward because the science upon which the regulator based his judgment was operating well within its power. But when the concern was subtle—how much cancer is caused by 10% of background radiation—science was being asked a question that lay beyond its power; the question was "trans-scientific." Yet the regulator, by law, was expected to regulate even though science could hardly help him: This is the regulator's dilemma.

This chapter was presented as a paper at the Workshop on De Minimis Risk; also at the National Academy of Engineering Symposium on Hazards: Technology and Fairness, Washington, D.C., June 3–4, 1985. A slightly modified version appeared in *Issues in Science and Technology* II (1):59–72, 1985.

Alvin Weinberg • Institute for Energy Analysis, Oak Ridge Associated Universities, Oak Ridge, Tennessee 37830.

Though my essay is subtitled "The Regulator's Dilemma," many of the same issues arise in the adjudication of disputes over who is to blame, and who is to be compensated, for damages allegedly caused by rare events. The regulator's dilemma is faced also by the toxic tort judge—indeed the regulator's dilemma could equally be called the "toxic tort dilemma."

A lawsuit involving alleged injury from chemical pollutants is unlike the traditional liability case. If my car injures a pedestrian, I am liable to be sued—but what is at issue is not whether or not I have injured the pedestrian; it is whether or not I am at fault in running into him. If the lead from my car's exhaust is alleged to cause bodily harm, the issue is not whether my car emitted lead but whether the lead actually caused the alleged harm. The two situations are quite different: In the first, the relation between cause and injury is not at issue; in the second, it is the issue.

In this chapter, therefore, I shall try to delineate more precisely those limits to science that give rise to the regulator's dilemma. I shall speculate on how these intrinsic limits to science seem to have catalyzed a profound attack on science by some sociologists and public-interest activists and I shall offer a few ideas that might help the harried regulators finesse these "trans-scientific" limits of science.

SCIENCE AND RARE EVENTS

Science deals with regularities in our experience; art deals with singularities. It is no wonder that science tends to lose its predictive or even explanatory power, when the phenomena it deals with are singular, unreproducible, and one of a kind—i.e., *rare*—rather than regular, reproducible, and with many instances. Though science can often analyze a rare event after the fact (say, the Cretaceous-Tertiary extinction), it has great difficulty predicting when such an uncommon event will occur.

I shall distinguish between two sorts of *rare* events—accidents and low-level sports. Accidents are large-scale malfunctions whose etiology is not in doubt, but whose *a priori* likelihood is very small. Three Mile Island and Bhopal are examples of accidents. The precursors to these events and the way in which the accidents unfolded are well understood. Estimates of the likelihood of the particular sequence of malfunctions is on less solid ground. As the number of individual accidents increases, prediction of their probability becomes more and more reliable. We can predict very well how many automobile fatalities will occur in 1986; we can hardly claim the same degree of reliability in predicting the number of serious reactor accidents in 1986.

Low-level sports are rare in a rather different sense than are accidents. We know that about 100 rads of radiation will double the mutation rate in a large population of exposed mice. How many mutations will occur in a population of mice exposed to 100 mr of radiation? Here the mutations, if induced at all by such low levels of exposure, are so rare that to unequivocally demonstrate an effect with 95% confidence would require the examination of many million mice. Though in principle this is not impossible, in practice it is. Moreover, even if we could perform so heroic a mouse experiment, the extrapolation of such findings to humans would still be fraught with uncertainty. Thus, the effects of very low-level insult in man are rare events whose frequency again is beyond the ability of science to predict with accuracy.

When dealing with events of this sort, science resorts to the language of probability—i.e., instead of saying that this accident will happen on that date, or that a particular person exposed to a low-level insult will suffer a particular fate, it tries to assign *probabilities* for such occurrences. Of course, where the number of instances are very large, or the underlying mechanisms are fully understood, the probabilities are themselves perfectly reliable. In quantum mechanics, there is no uncertainty as to the probability distributions. But in the class of phenomena we are speaking of here, even though the likelihood of an event's happening, or of a disease's being caused by a specific exposure, is given as a probability, *the probability itself is very uncertain.* One can think of a somewhat fuzzy demarcation between what I've called science and trans-science: the domain of science covers phenomena that are deterministic, or the probability of whose occurrence can itself be stated precisely; trans-science, the domain of events whose probability of occurrence is itself highly uncertain.

"SCIENTIFIC" APPROACHES TO RARE EVENTS

Despite the difficulties, science has devised mechanisms for estimating, however imperfectly, the probability of rare events. For accidents, the technique is probabilistic risk assessment (PRA); for low-level sports, a variety of empirical and theoretical approaches have been used.

Probabilistic Risk Assessment. Though probabilistic risk assessment had been used in the aerospace industry for a long time, it first sprang into public prominence with Professor Rasmussen's Reactor Safety Study, WASH-1400, which first appeared in 1975.[2]

Probabilistic risk assessment seeks to identify all sequences of subsystem failures that may lead to a failure of the overall system; it then tries to estimate the consequences of each system failure so identified. The output of a PRA is a probability distribution, $P(C)$; i.e., the probability, P, per reactor year, of consequence having magnitude C. Consequences include both material damage and health effects. Usually, the probability of accidents having large consequences is less than the probability of accidents having small consequences.

A probabilistic risk assessment for a reactor requires two separate estimates: first, an estimate of the probability of each accident sequence, and second an estimate of the consequences—particularly the damage to human health—caused by the uncontrolled effluents released in the accident. An accident sequence is a series of equipment or human malfunctions: a pump that fails to start, a valve that does not close, an operator confusing an "on" with an "off" signal. For many of these individual events, we have statistical data—i.e., enough valves have operated for enough years so that at least in principle we can make pretty good estimates of the probability of failure. Uncertainties still remain since we can never be certain that we have identified every relevant sequence. Proof of the adequacy of PRA must therefore await the accumulation of operating experience. For example, the median probability of a core melt in an LWR, according to the original Rasmussen report, was 5×10^{-5} per RY; the core melt at TMI-2 occurred after only 700 reactor-years. However, TMI-2 differed from the reactors treated by Rasmussen, and in retrospect, one could rationalize most of the discrepancy between the Rasmussen estimate and the seemingly premature occurrence at TMI-2.[3] Since TMI-2, the world's LWRs have accumulated some 1500 years of reactor operation without a core melt. This performance places an upper limit on the *a priori* estimate of the core-melt probability. Thus, if this probability were as high as 10^{-3} per RY (as had been suggested by Okrent[4]), then the likelihood of surviving 1500 reactor years would not be more than 22%; otherwise put, we can say with 78% confidence that the core-melt probability is not as high as 1 in 1000 reactor years. With 500 LWRs on line in the world, should we survive until 2000 without another core melt, we could then say with 95% confidence that the core-melt probability is not higher than 1 in 2000 reactor-years. In the absence of such experience, one is left with rather subjective judgments. Although the Lewis critique of Rasmussen's study[5] asserted that it could not place a bound on the uncertainty of PRA, Rasmussen has argued that his estimate of core-melt probability might be in error by about a factor of 10—that is, the probability may be as high as 1 in 2000 reactor-years or as low as 1 in 200,000 per RY. As we see, we can, after 3000 reactor years of operation without a core melt, say with about 78% confidence that Rasmussen's upper limit (1 in 2000 per RY) is not too optimistic. And if we survive to 2000 without a core melt, the confidence level with which we can make this assertion rises to 95%. Our confidence in probabilistic risk analysis can eventually be tested against actual, observable experience but until this experience has been accumulated, we must concede that any probability we predict must be highly uncertain. To this degree our science is incapable of dealing with rare accidents, but time—so to speak—annihilates uncertainty in estimates of accident probability.

Unfortunately, time does not annihilate uncertainties over consequences as unequivocally as it does frequency of accidents. A large reactor or chemical plant accident can cause both immediate, acute health effects and delayed, chronic effects. If the exposure either to radiation or to methyl isocyanate is high enough, the effect on health is quite certain. For example, a single exposure of about 400 rads will cause about half of those exposed to die. On the other hand, in a large accident there will also be many who are exposed to smaller doses, indeed to doses so low that the dose response is practically indeterminate. At Bhopal, 200,000 people were exposed to MIC and recovered. We cannot say positively whether or not they will suffer some chronic disability.

The very worst accident envisaged in the Rasmussen study, with a probability of 10^{-9} per RY, led to an estimated 3300 early fatalities, 45,000 early illness, and 1500 per year delayed cancers among 10^7 exposed people. Almost all of the estimated delayed cancers are attributed to exposures of less than 1000 milliroentgens per year—a level at which we are very hard put to estimate the risk of inducing cancer. Similarly, the American Physical Society's critique[6] of the Rasmussen study attributed an additional 10,000 deaths over 30 years among 10 million people exposed to Cs135 laid down in a very large accident. The average exposure in this case was 250 millirem per year, again a level at which our estimates of dose-response are extremely uncertain.

Has the nuclear community, particularly its regulators, figuratively shot itself in the foot by trying to estimate the number of delayed casualties as a result of these low-level exposures? In retrospect, I think the Rasmussen study would have been on more solid ground had it confined its estimates only to those health effects that resulted from exposures at higher levels, where science makes reliable estimates; for the lower exposures the consequences could have been stated simply as the number of man-rems of exposure of individuals whose total exposure did not exceed, say, 5000 mr, without trying to convert this number into numbers of latent cancers. Thus, health consequence would be reported in two categories: for highly exposed individuals, the number of health effects; for slightly exposed individuals, the total man-rems or even the distribution of exposures accrued by the large number of individuals so exposed. Perhaps some scheme such as this could be adopted in reporting the results of future probabilistic risk assessments: It at least has the virtue of being more faithful to the state of scientific knowledge than does the present convention.

LOW-LEVEL EXPOSURE

In both of my examples of accidents (Bhopal and reactors) many people are exposed to low-level insult; the uncertainties inherent in estimating the effects of such low-level exposure are heaped on top of uncertainties in estimating the probability of the accident that might lead to the exposure in the first place.

Science has exerted great effort to ascertain the shape of the dose-response curve at low doses—but very little, if anything, can be said with certainty about the low dose response. Thus, to quote the 1980 report (BEIR-III) of the National Academy of Sciences,"The Committee does not know whether dose rates of gamma or x-rays of about 100 mrads/yr are detrimental to man. . . . It is unlikely that carcinogenic and teratogenic effects of doses of low-LET radiation administered at this dose rate will be demonstrable in the foreseeable future."[7] All of which prompted President Handler to comment in his letter of transmittal to EPA, "It is not unusual for scientists to disagree . . . [and] . . . the sparser and less reliable the data base, the more opportunity for disagreement. . . . The report has been delayed . . . to permit time . . . to display all of the valid opinions rather than distribute a report that might create the false impression of a clear consensus where none exists."[7]

This forthright admission that science can say little about low-level insults I find admirable. It represents an improvement over the unjustified assertion in the BEIR-I report of 1972 that 170 millirems per year over 30 years, if imposed on the entire U.S. population, would cause between 3000 and 15,000 cancer deaths per year.[8] I do not quarrel with the estimated upper limit—which amounts to 1 cancer per 2500 man-rems; I regard the lower limit different from zero as being unjustified—and having caused great harm. The proper statement should have been that at 170 mr per year, we estimate the upper limit for the number of cancers to be 15,000 per year, and the lower limit might be zero.

Since the appearance of the BEIR reports, two other developments have added to the burden of those who must judge the carcinogenic hazard of low-level insults: (1) natural carcinogens and (2) ambiguous carcinogens.

Natural Carcinogens. Is cancer "environmental" in the sense of being caused by technology's effluents, or is cancer a natural consequence of aging? In the past few years I believe we have seen a remarkable shift in viewpoint: Whereas 15 years ago most cancer experts would have accepted a primarily environmental etiology for cancer, today the view that natural carcinogens are far more important than are manmade ones has gained many converts. Bruce Ames, in his famous *Science* article which was illustrated by Robert Indiana's modern painting *Eat-Die*, marshaled powerful evidence that many of our most common foods contain carcinogens.[9] Indeed, John Totter, supported by the late Philip Handler, has offered epidemiological evidence for the oxygen radical theory of carcinogenesis: that we grow older and eventually get cancer because we metabolize oxygen; and oxygen radicals can play havoc with our DNA.[10] As such views of the etiology of cancer acquire scientific support, I should think that the trans-scientific question "How much cancer is caused by a tiny chemical or physical insult?" will be recognized as irrelevant. One doesn't swat gnats in the face of a stampeding elephant.

Ambiguous Carcinogens. To further complicate the cancer picture, I call your attention to evidence that some agents, such as dioxin, various dyes, and even moderate levels of radiation, seem to diminish the incidence of some cancers at the same time they increase the incidence of others—so that the life-span of the treated animals on average exceeds that of the untreated ones.[11] A most striking example, given by Haseman, is yellow dye-14 given to leukemia-prone female F344 rats, which completely suppresses leukemia (which is always fatal) but causes liver tumors, most of which are benign.

I mention these two findings—or perhaps points of view—to stress my underlying point, that where we are concerned with low-level insult to human beings, we can say very little about the cancer dose-response curve. Saying that so many cancers will be caused by so much low-level exposure to so many people, a practice that terrifies many people, goes far beyond what science actually can say.

HOW SCIENCE REACTS TO INTRINSIC UNCERTAINTY

Does the scientific community accept the notion that there are intrinsic limits to what it can say about rare events: that as events become rarer, the uncertainty in the probability of occurrences of a rare event is bound to grow? Perhaps a better way of framing this question is: To what use can we put the tools of scientific investigation of rare events— say, probabilistic risk analysis and large-scale animal experiment as surrogates for epidemiological inquiry—if we concede that we can never get definitive answers?

For probabilistic risk analysis, I should say that an uncertainty as high as a factor of 10 is often useful, especially if one uses the PRA for comparing risks. For example, the 1500 reactor years already experienced since TMI suggests that a reactor core melt probability is likely to be less than 10^{-3} per year and may well be as low as PRA predicts, less than 10^{-4} per year. This is to be compared with dam failures whose probability, based on many hundreds of thousands of dam years (and where time has annihilated uncertainty), is around 10^{-4} per year. Even with this uncertainty, we can judge roughly how safe reactors are compared to dams.

When one compares the *relative* intrinsic safety of two very similar devices—like two water-moderated reactors—PRA is on much more solid ground. Here one is not asking for absolute estimates of risk, but rather estimates of relative safety. If the reactors, A and B, differ in only a few details—say reactor A has two auxiliary feed water trains whereas B has only one—the ratio of core-melt probabilities ought to be much more reliable than their absolute values, since the ratio requires an estimate of failure of a single subsystem, in this case, the extra AFW on reactor A.

Not only can one say with reasonable assurance how much safer reactor A is than reactor B; one can, as a result of the detailed analysis, identify the subsystems that contribute most to the estimated failure rate. Even if PRA is inaccurate, it is very useful in unearthing deficiencies: One can hardly deny that a reactor in which deficiencies revealed by PRA have been corrected is safer, even if one is unwilling to say how much safer.

Somewhat the same considerations apply to low-level insult. An agent that does not shorten life-span at higher dose will not shorten life-span at lower dose. An agent that is a very powerful carcinogen at high dose is more likely to be a carcinogen at low dose than one that is a less powerful high-dose carcinogen. Thus, animal experiments surely are useful in deciding which agents to worry about and which not to worry about. And of course the Ames test has made at least some preliminary screening of carcinogens more feasible. The difficulty today seems to be not so much identifying agents that at high dose may be carcinogens as it is prohibiting exposures far below levels at which no effect can be, or ever will be, demonstrated. The regulator and the concerned citizen are inclined to lean over backward so far as to approve the Delaney amendment, which forbids interstate commerce of any carcinogenic agent in food without ever saying anything about allowable levels or relative risks of, say, cancer induction by nitrosamines and digestive disorders caused by meat untreated with nitrates!

The Delaney amendment is the worst example of how a disregard of an intrinsic limit of science can lead to bad policy by overenthusiastic politicians. Harvey Brooks has often pointed out that one can never prove the impossibility of an event that is not forbidden by a law of nature; most will agree that a *perpetuum mobile* is impossible because it violates the laws of thermodynamics. That one molecule of PCB may cause a cancer in humans is a proposition that violates no law of nature: Hence, many, even within the scientific community, seem willing to believe that this possibility is something to worry about! It was this error that led to the Delaney amendment.

THE ATTACK ON SCIENCE FROM THE SOCIOLOGY OF KNOWLEDGE

When is an event so rare that the prediction of its occurrence forever lies outside the domain of science, i.e., within the domain of trans-science? Clearly we cannot say, and perhaps as science progresses, this boundary between science and trans-science will recede toward events of lower frequency. But at any state, the boundary is fuzzy, and much scientific controversy boils over deciding where that boundary lies. One need only read the violent exchange between professors Radford and Rossi over the risk of cancer from low levels of radiation to recognize that, where the facts are obscure, argument— even *ad hominem* argument—blossoms. Indeed, Alice Whittemore[12] has pointed out that at this "rare event" boundary between science and trans-science, facts and values are always intermingled. A scientist who believes that nuclear energy is evil because it inevitably leads to proliferation of nuclear weapons (which is a common basis for opposition to nuclear energy) is likely to judge the data on induction of leukemia from low-level exposures at Nagasaki differently than a scientist whose whole career has been devoted to making nuclear power work. Cognitive dissonance is all but unavoidable when the data are ambiguous and the social and political stakes are high.

No one would dispute that judgments of scientific truth are thus affected by the scientists' value system when the issues are at or close to the boundary between science and trans-science. On the other hand, as the matter under dispute moves away from that border into the domain of science, most would claim that the scientist's extrascientific values intrude less and less. Soviet scientists and American scientists may disagree on the effectiveness of a ballistic missile defense, but they agree on the cross-section of U235 or the lifetime of the pi-meson.

This all seems obvious, even trite. Yet in the past decade or so, a school of sociology of knowledge has sprung up in the United Kingdom which claims that "scientific views are determined by social (external) conditions, rather than by the internal logic of scientific tradition and inherent characteristics of the phenomenal world,"[13] or "all knowledge and knowledge claims are to be treated as being socially constructed: genesis, acceptance, and rejection of knowledge [is] sought in the domain of the Social World rather than . . . the Natural World."[14]

The attack here is not on science at the border, in particular, the prediction of the frequency of rare events. At least the more extreme of the sociologists of knowledge claim that the traditional ways of establishing scientific truth—by appealing to nature in a disciplined manner—is not how science really works even in situations very far from the science/trans-science border. Scientists are seen as competitors for prestige, for pay, for power, and it is the interplay between these conflicting aspirations, not the working of some underlying scientific ethic, that defines scientific "truth." To be sure, these attitudes toward science are not widely held by practicing scientists at the center of scientific activity; however, they are taken seriously by many political activists who, though not in the mainstream of science, nevertheless exert important influence on other institutions—the press, the media, the courts—that ultimately influence public attitudes toward science and its technologies.

If one takes such a caricature of science seriously, how can one trust an expert? If scientific truth, even at the core of science, is decided by negotiation between individuals in conflict because they hold different nonscientific beliefs, how can one say that this scientist's opinion is preferred to that one's? And if the matter at issue moves across the science/trans-science boundary, where all we can say with certainty is that uncertainties are very large, how much less able are we to distinguish between the expert and the charlatan, between the scientist who tries to adhere to the usual forms of scientific behavior and the scientist who suppresses facts that conflict with his political or social or moral preconceptions?

I don't think it will do to define a new branch of science, *regulatory science,* in which the norms of scientific proof are less demanding than are the norms in ordinary science. I should think that a far more honest and straightforward way of dealing with the intrinsic inability of science to predict the occurrence of rare events is to concede this limitation, and not to ask of science or scientists more than they are capable of providing. Regulators, instead of asking science for answers to unanswerable questions, ought to be content with less far-reaching answers. Where uncertainty bands can be established, they should regulate on the basis of uncertainty; where uncertainty bands are so wide as to be meaningless, recast the question so that regulation does not depend on answers to the unanswerable. And, since these same limits apply to litigation, the legal system ought, much more explicitly than it has heretofore, to recognize that science and scientists often have little to say, probably much less than some scientific activists would admit.

The *bona fides* of scientific adversaries often are at the heart of litigation over personal injury alleged to be caused by subtle, low-level exposures. Each side presents witnesses whose scientific credentials are regarded as impeccable by the side the witnesses are supporting. Since the issues themselves tend to be trans-scientific, one can hardly decide the validity of the "scientific" assertions of either side's witnesses. Under the circumstances, I suppose one is justified in regarding a scientific witness no differently than any other witness: His credibility is judged by his past record, behavior, and general demeanor, as well as the self-consistency of his testimony. Such, at least, was the way in which Judge Patrick Kelley settled the *Johnston* v. *United States* case, by impugning, on grounds no different from those one would invoke in an ordinary lawsuit, the competence if not the integrity of one side's scientific witnesses.

FINESSING UNCERTAINTY

Various approaches for finessing uncertainty can be identified. I shall briefly describe two of these—the technological fix and de minimis—without claiming that these are the most important, let alone the only ones.

Technological Fix. Science cannot predict exactly the probability of a serious accident in a light water reactor, or the likelihood that a radioactive waste canister in a depository will dissolve and release activity to the environment. Can one design reactors or waste cans for which the probability of such occurrences is zero—or at least which depend, to prevent such mishaps, on immutable laws of nature that can never fail rather than on incompletely reliable intervention of electromechanical devices? Surprisingly, this approach to nuclear safety has come into prominence only in the past 5 years. K. Hannerz[15] in Sweden and G. H. Lohnert[16] in Germany have each proposed reactor systems, PIUS and the modular high-temperature-gas-cooled reactor, whose safety depends not on active interventions but rather on passive, inherent characteristics. Though one cannot say that the probability of mischance has been reduced to zero, there is little doubt that the probabilities are several, perhaps three, orders of magnitude lower than the probabilities of mischance for existing reactors. To the extent that such reactors embody the principle of inherent safety, their adoption would avoid much of the hassle over reactor safety, Price-Anderson, repetition of Three Mile Island, and similar obstacles. In short, such a technical fix enables one largely to ignore the uncertainties in any prediction of core-melt probabilities.

The idea of incorporating *inherent* or passive safety in the design of chemical plants had been proposed, unbeknownst to the nuclear community, by Professor Theodore Kletz of the Loughborough University of Technology in 1984,[17] shortly after the disaster at the Flixborough cyclohexane plant, which killed 28 people. I should think that one of the main consequences of the Bhopal disaster will be incorporation of inherent safety into new chemical plants—again, a way of finessing uncertainty in predicting failure probabilities.

De Minimis. A perfect technical fix, such as a totally safe reactor, or a crash-proof car, is usually not available, at least at an affordable cost. Some low levels of exposure to materials that are toxic at high levels are inevitable, even though we can never accurately establish the risk of such exposures. One way of dealing with this situation is to invoke the principle of de minimis. This principle, as exposed by H. Adler and A. Weinberg,[18] argues that for insults that occur naturally and to which the biosphere has always been exposed, and presumably to which it has adapted, one should not worry about any additional manmade exposure *as long as the manmade exposure is small compared to the natural exposure.* The basic idea here is that the natural level of a ubiquitous exposure (like cosmic radiation), if it is deleterious, cannot have been very deleterious since in spite of its ubiquity, the race has survived. Moreover, we concede that we do not know, and can never know, what the residual effect of natural exposure really is. An additional exposure that is small compared to the natural background ought to be acceptable; at the very least, its deleterious effect, if any, can never be determined.

Adler suggested that for radiation whose natural background is well known, one might choose a de minimis level as the standard deviation of the natural background. This turns out to be around 20% of the mean background, around 20 mr per year, and this value has been used as the EPA standard for exposure to the entire radiochemical fuel cycle.

We know more about the natural incidence and about the biological effects of radiation than we do for any other agent. It would be natural therefore to use the standard established for radiation as a standard for other agents. This approach has been used by Professor Westermark of Sweden, who has suggested that for naturally occurring carcinogens such as arsenic, chromium, and beryllium, one might choose a de minimis to be, say, 10% of the natural background.[19]

Clearly, a de minimis level will always be somewhat arbitrary. Nevertheless, it seems to me that unless such a level is established, we shall forever be involved in fruitless arguments, the only beneficiary of which will be the toxic tort lawyers. Could the principle of de minimis be applied in litigation in much the same way it might be applied to regulation—i.e., if the exposure is below de minimis, then the blame is intrinsically unprovable and cannot be litigated? I would imagine that the legal de minimis might be set higher than the regulatory de minimis; for example, the legal de minimis for radiation might be the background (since the BEIR-III concedes there is no way of knowing whether or not such levels are deleterious). The regulatory de minimis could justifiably be lower, simply on grounds of erring on the side of safety.

One approach might be to concede that there is some level of exposure that is "beyond demonstrable effect" (BDE). This defines a trans-scientific threshold. A de minimis level might then be established at some fraction, say, one-tenth, of this BDE level. For example, if we take the previously quoted value of 100 mr per year of low-LET radiation as the BDE level for somatic effects, then a de minimis for low-LET might be set at 10 mr per year. Of course, such a procedure would evoke much controversy as to what is the BDE level, or whether 10 is an ample safety factor. This example demonstrates, however, that at least in the case of low-level radiation, a scientific committee was able to agree on a BDE level. As for the safety factor of 10, this cannot be adjudicated on scientific grounds. The most one can say is that tradition often supports a safety factor of 10—for example, the old standard for public exposure (500 mr per year) was set at one-tenth of the tolerance level for workers (5000 mr per year).

Can a principle of de minimis be applied to accidents? What I have in mind is the notion that accidents that are sufficiently rare might he regarded somehow in the same category as acts of God, and compensated accordingly. We already recognize that natural disasters should be compensated by the society as a whole. One can argue that an accident whose occurrence requires an exceedingly unlikely sequence of untoward events might also be regarded as an act of God. Thus the Price-Anderson Act might be modified so that, quite explicitly, accidents whose consequences exceeded a certain level, and whose probability as estimated by PRA would be less than, say, 10^{-9} per year, would be treated as acts of God. Compensation in excess of the amount stipulated in the revised act would be the responsibility of Congress. The cutoff for compensation or for probabilities would be negotiable, and perhaps would be revised every 10 years or so. One not entirely fanciful suggestion might be to set any probability of the order of 10^{-7} to 10^{-8} per year to be a de minimis cutoff, this being the frequency at which the earth may have been visited by the cometary asteroids that may have caused the geologic extinctions.

CONCLUSIONS

The reader must be aware that, as in most such questions, identifying and characterizing the problem is easier than solving it. That the regulator's and the toxic torts dilemma is rooted in science's inability to predict rare events cannot be denied. How to get the regulator and the toxic tort judge off the horns of the dilemma is far from easy, and my two suggestions are offered tentatively and with diffidence.

Equally obvious is the intrinsic social dimension of the issue. In an open litigious democracy such as ours, any regulation, any judicial decision can be appealed, and if the courts offer no redress, in principle, Congress can, but these mechanisms are ponderous. The result seems to me to be a gradual slowing of our technological-social engine—enmeshed more and more in fruitless argument over irresolvable questions.

Western society was debilitated once before by such fruitless tilting with Don Quixotian windmills. I refer of course to the devastating campaign against witches of the 14th to the early 17th centuries. As William Clark has put it so vividly, in this period society took for granted that death, disease, and crop failure could be caused by witches.[20] To avoid such catastrophes one had to burn the witches responsible for them—and some million innocent witches were burned as a result. Finally, in 1610, the Spanish inquisitor Alonzo Salazar y Frias realized there was no demonstrated connection between catastrophe and witches. Though he did not prohibit their burning, he did prohibit use of torture to extract confessions. The burning of witches and witch-hunting generally declined precipitously.

I have recounted this story many times by now. Yet it still seems to me to capture the essence of our dilemma: The connection between low-level insult and bodily harm is probably as difficult to prove as is the connection between witches and failed crops. That our society nevertheless has allowed this issue to emerge as a serious social concern I regard as an aberration, which in the modern context is hardly less fatuous than were the witch-hunts of the Middle Ages. That dark phase in Western society died out only after several centuries. I hope our open, democratic society can regain its sense of proportion far sooner and can get on with managing the many real problems we will always face rather than waste our energies on essentially insoluble, and by comparison, intrinsically unimportant, problems.

REFERENCES

1. W. D. Ruckelshaus, "Risk, Science, and Democracy," *Issues in Science and Technology 1*(3), 19–38, Spring 1985.
2. "Reactor Safety Study: An assessment of Accident Risk in U.S. Commercial Nuclear Plants," WASH-1400, NUREG 75/014, U.S. Nuclear Regulatory Commission, Washington, D.C. 1975.
3. N. Rasmussen, "Methods of Hazard Analysis and Nuclear Safety Engineering" in *Annals of the New York Academy of Sciences 365*, 29–33, April 24, 1981.
4. David Okrent, *Nuclear Reactor Safety, On the History of the Regulatory Process*, University of Wisconsin Press, Madison, Wisconsin, 1981.
5. "Risk Assessment Review Group Report to the U.S. Nuclear Regulatory Commission," H. W. Lewis, Chairman, NUREG/CR-0400, U.S. Nuclear Regulatory Commission, Washington, D.C., September 1978.
6. *Reviews of Modern Physics 47*, Supplement 1, 1975.
7. "The Effects on Populations of Exposure to Low Levels of Ionizing Radiation: 1980," BEIR-III, Committee on the Biological Effects of Ionizing Radiations, National Academy of Sciences, Washington, D.C., 1980.
8. "The Effects on Populations of Exposure to Low Levels of Ionizing Radiation," BEIR-I, Committee on the Biological Effects of Ionizing Radiations, National Academy of Sciences, Washington, D.C., 1972.
9. Bruce N. Ames, "Dietary Carcinogens and Anticarcinogens," *Science 221*(4617): 1249–1256, September 23, 1983.
10. John R. Totter, "Spontaneous Cancer and its Possible Relationship to Oxygen Metabolism," *Proceedings of the National Academy of Sciences 77*(4): 1763–1767, April 1980.
11. Alvin M. Weinberg and John B. Storer, "On 'Ambiguous' Carcinogens and Their Regulation," *Risk Analysis 5*(2): 151–155, June 1985.
12. Alice Whittemore, "Facts and Values in Risk Analysis for Environmental Toxicants," *Risk Analysis 3*(1): 23–33, March 1983.
13. J. Ben-David, "Emergence of National Traditions in the Sociology of Science, The United States and Great Britain," *Social Inquiry 48*(3–4): 197–218, 1978.
14. Trevor J. Pinch and Wiebe E. Bijker, "The Social Construction of Facts and Artefacts: or How the Sociology of Science and the Sociology of Technology Might Benefit Each Other," *Social Studies of Science 14*: 399–441, 1984.
15. K. Hannerz, "Towards Intrinsically Safe Light Water Reactors," Oak Ridge Associated Universities, Institute for Energy Analysis, Oak Ridge, Tennessee, ORAU/IEA-83-2(M) Rev., June 1983.
16. Herbert Reutler and Gunther H. Lohnert, "The Modular High Temperature Reactor," *Nuclear Technology 62*: 22–30, July 1983.
17. Trevor A. Kletz, *Cheaper, Safer Plants or Wealth and Safety at Work—Notes on Inherently Safer and Simpler Plants*, The Institution of Chemical Engineers, Rugby, England, 1984.
18. H.I. Adler and A.M. Weinberg, "An Approach to Setting Radiation Standards," *Health Physics 34*: 719–720, June 1978.
19. T. Westermark, "Persistent Genotoxic Wastes—An Attempt at a Risk Assessment," Royal Institute of Technology, Stockholm, Sweden, 1980.
20. William C. Clark, "Witches, Floods, and Wonder Drugs: Historical Perspectives on Risk Management," RR-81-3, International Institute for Applied Systems Analysis, Laxenburg, Austria, March 1981.

II

Quantitative Aspects of De Minimis Risk

5

Significant Risk Is Not the Antonym of De Minimis Risk

Daniel Byrd III and Lester Lave

INTRODUCTION

Recent events have led to an increased interest in a "significant risk" policy for the regulation of exposures to toxic chemicals. In 1980 the Occupational Safety and Health Administration (OSHA) promulgated a rule in which the standard for any carcinogen was to be lowered to the extent technologically and economically feasible.[1] In a 1980 decision, the Supreme Court vacated a rule that established a standard for exposure to benzene, based on a statutory interpretation, although the Court did not overturn OSHA's entire cancer policy.[2] An agency had to make a finding that a risk was *significant* before it could consider regulating it, and the finding had to be part of the record.

The "de minimis" concept from common law holds that the court does not concern itself with trivia. Logically, a finding of de minimis risk would be sufficient to conclude that an exposure was not a significant risk and not of concern. The converse is not necessarily true. A risk that is not de minimis still may not be significant.

The standard of economic *feasibility* for OSHA that was upheld by the Supreme Court in the cotton dust case is that an entire industry cannot be eliminated, but a few participants can be put out of business.[3] The first standard would bankrupt the weakest firms, the second the next weakest firms, and so on, until enough standards had been promulgated to eliminate most firms. Clearly, any single-minded pursuit of the notion

The views expressed in this chapter are those of the authors and do not necessarily represent the official views of the Halogenated Solvents Industry Alliance or Carnegie-Mellon University.

Daniel Byrd III • Halogenated Solvents Industry Alliance, Washington, D.C. 20036. *Lester Lave* • Graduate School of Industrial Administration, Carnegie-Mellon University, Pittsburgh, Pennsylvania 15213.

that risks are to be lowered to the extent economically feasible would impoverish the nation, with each industry pushed to the edge of bankruptcy (assuming an agency did its economic feasibility analysis accurately). More important, a well-informed OSHA would remove any "surplus profit" in its first few regulations and leave the firms with no resources to tackle later, possibly more difficult, problems as they were discovered. If an industry is subject to more than one standard, then presumably most firms could be eliminated. From an economics perspective, the Court made a correct decision in the benzene case.

In interpreting the significant risk policy liberally, an agency cannot impose a more stringent regulation than one that reduces the risk below a nonsignificant level, just as it cannot take up the regulation of an insignificant risk. While the ruling forced the agencies to pause and think, it has not helped them arrive at working definitions of significant risk. Indeed, we will argue that, while some guidance is possible, ultimately no simple definition is possible.

In the short term the use of a significant risk policy could provide a cornerstone to improve regulatory risk decisions. It could promote consistency and would help allocate an agency's resources to high priority risks. Unfortunately, we believe that the policy is flawed conceptually. It cannot bear the weight being placed on it and must give way to a more satisfactory notion, which we propose later.

In the following discussion, we begin with the current world of the regulatory agencies, what the significant risk concept has to contribute pragmatically, and how it differs from de minimis risk. We compare the decisions of various agencies, as well as the decision rules of what constitutes a significant risk. We illustrate some fundamental flaws in the significant risk policy and then pose an alternative with superior properties.

STATUTES AND INTERPRETATION OF SIGNIFICANT RISK BY THE COURTS

"Significant" is not the antonym of "de minimis" in the assessment of risks to health or the environment. The two policies are neither linked semantically nor are they dichotomous. They arose at different times in the evolution of environmental regulation. They have different philosophical underpinnings. One does not begin where the other ends. For this reason, when both significant and de minimis risks are described in terms of magnitude, we suggest that the term "nontrivial but insignificant" (NTI) is useful to describe the region of risk that lies between them.

In addition, the term "insignificant" is not used in environmental statutes, nor is it a synonym for de minimis. Instead, the idea of a de minimis risk policy originates from a concept of what is socially trivial. While agencies have been reluctant to commit themselves to definitions of these terms, particularly in view of expected legal challenges and adverse publicity, at least three agencies have tested policies related to risk magnitudes.

The Food and Drug Administration (FDA) declared that a food additive with a contaminant that poses less than a one-in-a-million lifetime risk (10^{-6}) will not be treated as a carcinogen under the Delaney clause.[4] FDA apparently believes that risks below this administrative threshold can be exempted from consideration as trivial. In FDA's view a 10^{-6} risk is not de minimis, since de minimis risk was posed as an alternative concept in promulgating the policy, but such a risk may be insignificant.[5] The Environmental Protection Agency (EPA) proposed to regulate the level of the pesticide Larvadex in chickens and eggs such that the lifetime individual risk of cancer would be less than 10^{-6} in the consuming population.[6] The Nuclear Regulatory Commission (NRC) also has set quantitative safety goals. The NRC has defined these safety goals for nuclear power plants as no more than a 0.1% increase in cancers in the surrounding population over those normally prevailing without the reactor or a 0.1% increase in immediate deaths in the surrounding population.[7] The NRC goals are about 42×10^{-6} for immediate deaths and 180×10^{-6} for cancer deaths (per lifetime). The NRC goals might be interpreted as nonsignificant. Thus, when plant risks are at or below this level, they are of no practical concern.

The FDA interprets its 10^{-6} criterion cautiously. For any single additive, it represents a lower risk level than the NRC criterion, which in turn has seldom been translated into reactor licensing decisions. EPA withdrew the Larvadex proposal because of new data about its risk; this was not necessarily a policy change. Other agencies have not been this bold.

"Significant risk" is a frequently used, but judgmental term. It is an older policy that implies a requirement for more powerful information to act. As currently implemented in six federal statutes and 25 regulations, it was meant to give freedom and discretion to the individual making a decision.[8,9] The framers of these statutes and regulations do not appear to have had some quantitative definition in mind. Significant risk was something you would know when you confronted it. The authors probably believed that the notion of significant risk would change from individual to individual and from time to time.

A significant risk policy also has been applied in the interpretation of legislation that does not use the exact term, as described below. In this section, we focus on sections of three public health laws of particular interest to us. While unmeasurable risks may be viewed as significant in some circumstances, we arrive at a working definition of significant risks as measurable, hence capable of being observed.

Section 4(f) of the Toxic Substances Control Act of 1976 (TSCA) (8b) states that EPA must take action when "there may be a reasonable basis to conclude that a chemical substance or mixture presents or will present a *significant risk*" of serious or widespread harm to humans from cancer, gene mutations or birth defects. . . (emphasis added).

The Occupational Safety and Health Act of 1970 (OSH Act)[10] does not use the term "significant" to describe risks. Instead, the Supreme Court applied this policy in interpreting the relationship between Sections 3(8) and 6(b) of the OSH Act, which do not even refer directly to risk but which state, respectively, that the Occupational Safety and Health Administration (OSHA) should "promulgate, modify or revoke any occupational safety or health standard . . ." that is "reasonably necessary or appropriate to provide safe or healthful employment or places of employment. . . ." The Court reached three important conclusions that (1) these sections in conjunction require OSHA to apply a significant risk standard before undertaking development of a regulation, (2) the threshold finding for significant risk is defined in terms of risk magnitude, and (3) the risk should be characterized in quantitative terms sufficiently that the secretary of labor could understand the significance.[2]

The Clean Air Act (CAA) of 1970 (including the amendments of 1977) states in Section 112 that EPA should regulate a substance that "causes or contributes to air pollution which may reasonably be anticipated to result in an increase in mortality or an increase in serious irreversible or incapacitative reversible illness. . . ."[11] Like the OSHAct, Section 112 does not use the term "significant" to describe the risk. Instead, the courts have so far interpreted this section in terms of significant risk.[12] We realize, however, that this interpretation is controversial and that either current litigation or pending legislation may change how the CAA is to be implemented.

Significant risk policies have been widely applied and interpreted. For example, in the Federal Mine Safety and Health Regulatory Commission decision of April 7, 1981, the courts stated that "we hold that a violation is of such a nature as could significantly and substantially contribute to the cause and effect of a mine safety or health hazard if, based upon the particular facts surrounding that violation, there exists a reasonable likelihood that the hazard contributed to will result in an injury or illness of a reasonably serious nature. . . ."[13] We suspect that the definitions and experience of significant risk policies apply even more widely in that officials who implement federal statutes watch for developments in other areas regarding their responsibilities.

We do not believe that the term "significant risk" must be left as a vague concept. Instead, we think that the concept of significant risk can be based on the notion of an observable or measurable event. We propose as a definition that the significant event is potentially detectable amidst the confounding events. Certainly, it seems a paradox to state that a risk is of significant magnitude but too small to be detected. While we admit that this paradox can occur, we will illustrate how significant risks can be defined as detectable, nonsignificant risks as undetectable, and de minimis risks still lower than nonsignificant.

SOME FACTORS IN THE DESCRIPTION OF RISK

To have a significant risk policy based on detectability, some agreement must be reached about the conventions used to calculate and express risk. In this section, we suggest some definitions and explore their consequences. As we shall see, three concepts (type of hazard, population size, and probability of harm) go into the definition of significant risk.

A "hazard" is an undesirable event that might occur, such as an injury to health. A "risk" can be either the probability of this event occurring to an individual in a population or of the number of times the event could occur in a population. These terms are not used uniformly, and often one hears both that the risk of highway transport is 45,000 deaths per year and that the risk of being killed in a highway mishap is 1 in 4000 per year for the average American. When risk refers to the expected number of mishaps in a population, it is the product of the size of the population at risk and the probability that an individual will experience the hazard.

Some hazards are more feared than others: Death in an auto crash is viewed as less fearsome than death due to cancer. Becoming a quadraplegic is viewed by some as worse than death. Death to children is viewed as worse than death to adults. These views are not fixed but rather change, depending on the group involved, experience over time, and personal involvement. There also seems to be a stronger reaction to the death of several individuals together, such as the loss of a family, than to the loss of each individual in the unit at different times. When 350 highway deaths occur during a three-day period, this is an unremarkable, expected series of misfortunes. When they occur in a jumbo jet crash, this makes national headlines and leads to demands for fixing the air transport system.

For now, we gloss over these subtleties and define our goal as an improvement in U.S. health statistics. A significant risk policy is the tool for achieving the goal. For mortality, the goal will lead to a focus on the U.S. life table.[14] A life table is a tabular summary of the number of deaths among persons at each age in the population. Probability statements of risk (or expected numbers of deaths occurring among persons of average age in the population) can be converted to loss of life expectancy statements by reference to the U.S. life table, together with some assumptions.[15]

The information on which an assessment is based, however, often does not relate directly to a human life table. Typically, animal toxicology data are used. Various linear hazard functions currently are used to interpolate animal toxicology data between different environmental exposures.[16] They generally assume that the incidence of a health effect is proportional to exposure level; that is, the effect is qualitative (all or none), and the incidence changes with exposure, but the severity does not. The number of expected cancer cases is taken into account, but the kind of cancer is not. This assumption ordinarily is believed to apply to genotoxic events, mainly the processes of carcinogenesis and mutagenesis, but it seldom is tested against the data. Further, the linear models are described as giving an upper bound to potency, not a best estimate. This applies equally to animal and human extrapolation. The rationale for the upper-bound nature of the estimates is primarily that all "true" dose-response curves will give lower incidence estimates than a linear model under almost all cases. Furthermore, not all cancers are the same, since some are easily treated and rarely are fatal (like basal cell skin cancer), whereas others have low survival rates (like lung cancer). Thus, the linear model in conjunction with counting all cancers equally gives an upper bound to potency.

We focus on carcinogenic risks, since these models have been explored most and make our point most clearly. Examination of only one effect (cancer) also has the desirable effect of simplifying the aggregation of outcomes. For example, to add terata and cancers would require some kind of interconversion factor.

A hazard estimate is combined with an individual's exposure to obtain an expected individual risk. Usually, point estimates of exposure and an upper bound to potency are multiplied together. It also is useful to confine exposure estimates to a single substance, rather than many. One problem that creates confusion about significant risk is that detection of exposure arises from techniques of analytical chemistry that are not linked to the factors that create risk. For substance A the techniques may permit highly sensitive detection and accurate measurement, but the intrinsic risks may be low (e.g., food additives) whereas for substance B the situation may be reversed (e.g., coke oven emissions). The same is true of the precision in the measurement of exposure. Knowledge of the variation in measuring background levels (as opposed to the levels themselves) does not directly contribute to our knowledge of the background variation in risks from a substance.

From our studies, we know that a few individuals are exposed to high concentrations, a larger number to a moderate concentration, and the vast majority to a tiny concentration. These exposures generally can be approximated by a log-normal distribution. For some substances, however, this assumption clearly is untenable. For example, higher exposures in the workplace may result in a multimodal distribution. The primary difficulty in making point assessments of risk is the arbitrary selection of one exposure. Conceptually, the median exposure in a log-normal distribution is not a good indication of the exposure of the group at greatest risk, and it clearly does not protect the most exposed group, although lowering the exposure to all exposed will lower the median.

For an acute hazard, use of an arbitrary "most exposed" person also leads to problems. An extra margin of safety typically is factored into any standard derived from the higher exposure; this varies from substance to substance, because the variance from the mean differs.[17] For lifetime exposures, which are appropriate with chronic hazards, an individual accumulates many daily exposures. The average is a more appropriate value for risk calculations, unless it can be shown that a subpopulation has a higher average lifetime exposure for some reason, such as proximity to a production plant.

Another factor in the description of risk is the kinetics of exposure. The times of onset and duration of exposure to a chemical substance modify the impact on the life table. In addition, if the exposed population is not of average age, or if a latency period occurs between an exposure and disease, a constant risk may not translate into a constant loss of life expectancy. In the calculations below, we have finessed this problem by assuming a steady-state exposure.

Exposure and population are interconnected subjects. The population at risk and the population of concern are not necessarily the same. If a highly exposed but small subpopulation exists, then individual risk for each member of this group (the product of exposure and hazard—the expected incidence of cancer from the specific exposure) may be high. A large portion of the subpopulation may experience cancer from this cause. When these cases are considered in relation to the entire population, however, the average individual risk can be much lower. The significance of observations of cancer cases would be made by reference to the U.S. life table. No particular reason exists to treat the expectations for different persons differently.

Hazard, exposure, and population may not be independent variables, in which case risk is a complex function of the three factors. Each constituent factor can be described as a probability distribution, which in aggregate would give a more precise description of the uncertainty in the expectation.

On top of all other problems, we have very little guidance from the data on what actually is going on in nature. Measures of uncertainty in data quality and ways to incorporate the measures into estimates of risk need development. In the absence of a stated probability distribution, it is hard to dismiss a point estimate of risk that is based on poor quality information. It also is hard to take regulatory action based on a very hypothetical risk estimate, since the courts may overturn the regulation. For example, estimates of the risk from radiation have greater certitude than do ones for chemical substances. Linear damage functions for some sources of radiation can be justified as accurate, rather than upper-bound estimates, as is the case for chemical substances. For other radiation sources, linear-quadratic functions are thought to give accurate estimates.[18]

RISK CALCULATIONS

A regulatory agency could give a working definition of "significant risk" by declaring its practices about the factors that contribute to risk (perhaps in the form of guidelines) and providing the formulae to calculate risk. Bearing in mind that it is important to state whether the time period is annual or lifetime, we have illustrated some examples, using the following nomenclature:

S_x = an indicator (when positive) of the existence of significant risk from a substance, C_x = a constant that indicates the level of significant risk, selected as a matter of policy, x = an indication of the breadth of applicability, H = highest expected incidence at a specified exposure, E = expected exposure, I = individual risk, P_e = number of people in the population at risk (exposed), P_T = total population examined (of concern), A,B = arbitrary constants to adjust nonlinear equations for boundary conditions.

1. *Individual risk* is an upper bound to relative risk for the exposed person, calculated as follows:

$$S_i = I - C_i = (H \times E) - C_{individual} \tag{1}$$

2. The number of *persons affected* is some absolute number of persons in a population of any size, calculated as:

$$S_p = (I \times P_e) - C_{persons} \tag{2}$$

3. *Population risk* is the number of affected persons per size of population of concern (or aggregate risk), calculated as follows:

$$S_c = (I \times P_e \, P_T) - C_{concern} \tag{3}$$

For example, EPA's pending regulation of acrylonitrile under section 112 of the CAA is a current test of how to determine the population at risk. Most acrylonitrile comes from plants in a few states. If the population of concern is identified as the population of these states, the risk from these plants may merit attention that would not be appropriate when the denominator of the risk calculation is increased to the total U.S. population.

4. Nonlinear functions can be used. An infinite number of forms is possible, and even simple functional forms will rapidly proliferate. As an example, a "sliding" function of the number of persons exposed can be used to address the problem of potentially inconsistent risk decisions with different size populations. All other things being equal, a risk of 10^{-3} in a population of 10^4 is less hypothetical (and of greater concern) than a risk of 10^{-7} in a population of 10^8. Both situations imply that 10 people will be affected, but the relative uncertainty is greater for a risk of 10^{-7}. By applying an exponent of less than one to the population size, some allowance can be made for this difference. Milvy used a power function to adjust population measures.[19] A statement of risk significance with a power function of population will cover a wider range of situations, as follows:

$$S_n = (I \times A \times P_e{}^B) - C_{nonlinear} \tag{4}$$

More "sophisticated" equations, such as nonlinear hazard functions or functions that refer to existing background rates, also can be used. If multiple nonlinear terms are used, they will "interact" in unexpected and nonintuitive ways. This problem suggests that the use of linear terms in risk equations is largely a convenience.

NUMERICAL CRITERIA FOR SIGNIFICANT RISK

Since our definition of significant risk depends on an observable event, the first place to look for guidance is the life table for U.S. citizens.[14] While risk magnitude alone is simplistic, to obtain some idea of a criterion for risk significance, a reasonable approach might be to look at the ability to detect risk in the lowest risk component of the population. If an expected risk could be verified in principle by measurement, it would be significant.

What increase in mortality constitutes a significant increase will depend on sample size, control for confounding factors, and the commonness of the cause of death. However, one rule of thumb from epidemiology is that a relative risk must be 2 or greater to be significant; i.e., the risk of death from this cause must be doubled. For mortality from all causes, the smallest detectable increase in mortality would be for 11- to 12-year-old girls, since their death rate is about 2.4×10^{-4} per year. However, a much smaller increase in cancer could be detected, since its incidence is so much lower among this group. A carcinogen that causes less than an additional 3×10^{-5} cancer risk in 11- to 12-year-old girls is undetectable. Conceptually, a carcinogen causing more than this increase in total cancers in this group could be detected. This increase in risk is also a minimum for the entire population, since the statistical power to detect a risk for the most sensitive component of the population must be at least as great as that for the entire population.

In Chapter 7, Milvy has suggested that individual and population risk are related to each other in terms of the square root of the population size, rather than the number of individuals, at risk. We suggest that this relationship might be the case because the uncertainty in counting objects is proportional to the square root of the objects counted with the Poisson distribution, which is an appropriate distribution to use when looking for rare, random, all-or-none events in a population, as is the situation here. The Poisson distribution also is consistent with linear interpolation of hazard.

One way to view the relationship between individual and population risks is to relate the inherent uncertainty in counting events to the population under observation. Thus, for 10^4 cancer victims, even if the estimation were perfect, the minimum variability would be 10^2 to be sure with a confidence limit of 1 standard deviation that risk had changed. This approach suggests, then, that the minimum change in the level of risk that can be termed significant owing to measurability in a population of 10^4 is 100 out of 10,000 $(10^2/10^4)$ or 10^{-2}. For a population of 10^6, a similar approach leads to significant risk of 10^{-3} $(10^3/10^6)$, or for a population of 10^8 a risk of 10^{-4} $(10^4/10^8)$.

The criterion of observability also suggests that at least one person must die in a group to be significant. Thus, the product of group size and risk level define significance. For a tiny group, the excess risk would have to be very large to be significant. For a large group, significance would be determined by the square root of the expected number of victims.

Cohen has suggested that many people tend to treat risks, in the range from about 10^{-3} to 10^{-9}, equivalently.[15] Changes below these levels might not be perceived as measurable by many persons. While little evidence exists to support this view of risk perception, if it occurs, then a risk above 10^{-3} clearly is significant, even to them. The Supreme Court also has suggested roughly 10^{-3} as a risk level from occupational exposures that a reasonable person would consider significant.[2] These two proposals raise the possibility of a hypothetical risk perception threshold.

Since key individuals make agency decisions, it is reasonable to expect that agency regulatory decisions might reflect their individual risk perception. We proceed next to compare the theoretical benchmarks for risk significance to regulatory agency perceptions, as evidenced by agency practices.

AGENCY PRACTICES

With some assumptions, boundaries can be placed on the benchmark level of significant risk (C_x) in use at an agency, by observing its practices. In essence, we solved Equations 1 and 2 for C_x with $S_x = 0$ with data from published and proposed regulations. We give the results in Tables 1 and 2, as the apparent working definitions of what gets regulated or not, under the three statutes that interest us in this discussion. Consistent with the logic developed earlier we have used linear, upper-bound estimates, and the highest such estimate, when several are available.

Exposed populations often are not stated in agency publications. We obtained this information from support documents or guessed the appropriate number by comparison of other data. For each equation, we developed two categories. The "before regulation" category describes what we think was an agency's expectation of the risk at the time a decision was made to regulate—thus, that a significant risk existed. The data are "censored;" that is, the agency benchmark lies below this level. Part of the other category is taken from statements about substances that were not regulated. However, agencies seldom take the trouble to supply a rationale for actions not taken. We have supplemented the second category with statements about residual risks after regulation (either as proposed or in a final rule, whichever came last), bearing in mind that the agency still might regard this level as significant. Feasibility considerations may have led to the level of residual risk.

None of the agencies provide a population of concern. Therefore, we have not calculated risk to the population of concern. Presumably, if these data were available, a better differentiated profile for the individual organizations would emerge. Estimates of exposure were particularly difficult to obtain for OSHA regulations, because that agency uses the standard as a surrogate exposure. Little information is provided regarding the actual exposure expected after a particular standard is in place. Since we are interested primarily in internal agency perceptions, we have used the information provided by OSHA, namely, the standard. This may be exactly the point; it is agency perception that matters, not reality. For this and similar reasons, the risk levels from regulations for different substances are not strictly analogous, even across the same organization. We have made comparisons for the purpose of illustrating hypothetical decision making. Reference risks discussed in the previous section have been included as benchmarks in Tables 1 and 2. We have arbitrarily listed risks of less than 10^{-6} (lifetime individual) and less than 0.5

Table 1. Upper Limit of Lifetime Excess Individual Risk from Selected Regulated and Unregulated Substances

Risk	Substance (statute; reference)
Perceived level before regulation of a significant risk	
4×10^{-1}	Arsenic, workers (OSHA[20])
2×10^{-1}	Ethylene dibromide, workers OSHA[21])
1×10^{-1}	Ethylene oxide, workers (OSHA[22])
6×10^{-2}	Asbestos, workers (OSHA[23])
3×10^{-2}	Arsenic, primary copper (CAA[30])
2×10^{-2}	Coke oven emissions (CAA[24])
$>1 \times 10^{-2}$	Methylenedianiline (TSCA[25])
$>1 \times 10^{-2}$	Butadiene (TSCA[26])
1×10^{-2}	Uranium mines (CAA[27])
5×10^{-3}	Benzene, coke ovens (CAA[28])
2×10^{-3}	Benzene, fugitive emissions (CAA[29])
1×10^{-3}	Hypothetical Risk Perception Threshold (p. 49)
1×10^{-3}	Radionuclides, phosphate mines (CAA[27])
8×10^{-4}	Arsenic, glass manufacture (CAA[30])
8×10^{-4}	Radionuclides, DOE (CAA[27])
2×10^{-4}	Least U.S. Total Mortality Risk (p. 48)
2×10^{-4}	Coke ovens, workers (OSHA[32])
1×10^{-4}	Radionuclides, NRC licensees (CAA[27])
3×10^{-5}	Least U.S. Cancer Mortality Risk (p. 48)
1×10^{-6}	Potential De Minimis Risk (see p. 49)
Perceived risk levels	
2×10^{-2}	Asbestos, workers (OSHA[23])[a]
8×10^{-3}	Arsenic, workers (OSHA[20])[a]
8×10^{-3}	Ethylene dibromide, workers (OSHA[21])[a]
3×10^{-3}	Ethylene oxide, workers (OSHA[22])[a]
2×10^{-3}	Arsenic, primary lead smelters (CAA[30])[b]
1×10^{-3}	Hypothetical Risk Perception Threshold (p. 49)
1×10^{-3}	Formaldehyde, abrasive manufacturers (TSCA[30])[b]
1×10^{-3}	Arsenic, zinc oxide plants (CAA[30])[b]
2×10^{-4}	Least U.S. Total Mortality Risk (p. 48)
6×10^{-4}	Formaldehyde, apparel workers (TSCA[31])[b]
6×10^{-4}	Formaldehyde, funeral workers (TSCA[31])[b]
5×10^{-4}	Benzene, fugitive emissions (CAA[29])[a]
4×10^{-4}	Benzene, storage vessels (CAA[33])[b]
4×10^{-4}	Arsenic, secondary lead smelters (CAA[30])[b]
1×10^{-4}	Formaldehyde, conventional homes (TSCA[31])[b]
1×10^{-4}	Formaldehyde, mobile homes (TSCA[31])[b]
1×10^{-4}	Benzene, ethylbenzene plants (CAA[33])[b]
8×10^{-5}	Benzene, maleic anhydride plants (CAA[33])[b]
3×10^{-5}	Least U.S. Cancer Mortality Risk (p. 48)
3×10^{-6}	Formaldehyde, college students (TSCA[31])[b]
1×10^{-6}	Potential De Minimis Risk (p. 49)

[a] Postregulation risk level.
[b] Risk level not significant.

Table 2. Upper Limit of Persons Affected Annually by Selected Regulated and Unregulated Substances

Persons	Substance (statute; reference)
Perceived numbers of persons affected before regulation of a significant risk	
17,000	Asbestos, workers (OSHA[20])
>11,000	Formaldehyde, conventional homes (TSCA[31])
10,000	Minimum Variance in 10^8 Population (p. 48)
600	Formaldehyde, mobile homes (TSCA[31])
400	Formaldehyde, apparel workers (TSCA[31])
300	Ethylene dibromide, workers (OSHA[21])
300	Ethylene oxide, workers (OSHA[22])
100	Minimum Variance in 10^4 Population (p. 48)
11	Vinyl chloride (CAA[34])
9	Coke oven emissions (CAA[24])
6	Uranium mines (CAA[27])
5	Benzene, coke ovens (CAA[28])
3	Arsenic, primary copper (CAA[30])
0.5	Benzene, fugitive emissions (CAA[29])
0.5	Potential De Minimis Risk (p. 52)
0.3	Arsenic, glass manufacture (CAA[30])
0.07	Radionuclides, DOE (CAA[27])
0.05	Radionuclides, phosphate mines (CAA[27])
0.001	Radionuclides, NRC licensees (CAA[27])
Unstated	Coke ovens, workers (OSHA[32])
Unstated	Arsenic, workers (OSHA[20])
Perceived numbers of persons affected	
11,000	Asbestos, workers (OSHA[23])[a]
10,000	Minimum Variance in 10^8 Population (p. 48)
100	Minimum Variance in 10^4 Population (p. 48)
40	Formaldehyde, funeral workers (TSCA[31])[b]
>20	Methylenedianiline (TSCA[25])[b]
>20	Butadiene (TSCA[26])[b]
20	Ethylene dibromide, workers (OSHA[21])[a]
10	Formaldehyde, college students (TSCA[31])[b]
8	Ethylene oxide, workers (OSHA[22])[a]
7	Formaldehyde, abrasive manufacturers (TSCA[31])[b]
0.6	Vinyl chloride (CAA[34])[a]
0.5	Potential De Minimis Risk (p. 52)
0.4	Arsenic, secondary lead smelters (CAA[30])[b]
0.1	Benzene, fugitive emissions (CAA[29])[a]
0.08	Arsenic, zinc oxide plants (CAA[30])[b]
0.07	Arsenic, primary lead smelters (CAA[30])[b]
0.04	Benzene, storage vessels (CAA[33])[b]
0.03	Benzene, maleic anhydride plants (CAA[33])[b]
0.006	Benzene, ethylbenzene plants (CAA[33])[b]

[a] Postregulation risk level.
[b] Risk level not significant.

persons (annually in a population of 500,000 to 5,000,000) as de minimis levels, for purposes of illustration.

SOME ADVANTAGES OF SIGNIFICANT RISK

The information in the tables makes it clear that agencies act to decrease risks, but higher risks get reduced more often and to a greater extent. As risks get lower, so does risk reduction. Actions taken under the "significant risk" clauses of statutes are consistent with the risk magnitudes described in the tables. The range of scatter, however, is wide.

We have compared the significant risk concept to the de minimis concept. For example, benchmarks have been included for de minimis risk of less than 10^{-6} individual lifetime risks or fewer then 0.5 persons affected annually. A lifetime risk of 10^{-6} corresponds to about six seconds out of an average life-span. It is impossible to enumerate less than one person affected in a population.

The magnitudes of some of the risks thought not to merit regulation differ from the risk magnitudes associated with de minimis risk levels. None of the substances that merited analysis as significant risks under the three statutes caused a maximum estimated level of individual risk less than de minimis, either before or after regulation. From the perspective of persons potentially affected, the situation is somewhat different. A substantial portion of the substances, either pending regulation or after regulation, might affect more than 0.5 persons. Nevertheless, all of the substances proposed for regulation do meet one of the tests for significant risk, calculated as either individual risk or persons affected, except for the regulation of radionuclides for NRC licensees under the CAA. De minimis risk levels do not explain the behavior of these agencies as well as do significant risk levels.

Significant risk is not just the opposite of de minimis risk. As we have illustrated, a gap (the NTI region) exists between individual risk and the number of persons affected under significant risk policy versus de minimis one. A significant risk is compelling, whereas a de minimis risk is trivial. A nontrivial risk is not necessarily significant. For any given substance, the range of risk between these two levels corresponds to maximum uncertainty in a decision to take action. The risk levels do not clearly warrant action or dismissal.

From the perspective of a regulatory agency attempting to explain its actions, a significant risk policy has advantages over a de minimis one. Because a significant risk is more likely to fall in the range where severity also changes with dose, the need to assume an all-or-none risk is less important. Similarly, because the level of significant risk will arise from an exposure that is closer to the exposure in the experimental (or epidemiological) studies used to calculate the risk, significant risks will be less hypothetical than de minimis risk. With a significant risk policy, competing risks will be less likely to overwhelm the impact of the risk on life expectancy. When animal studies form the basis of the hazard estimate, corroborative epidemiological studies can make an important contribution. Potential detectability is a rationale for action, whereas triviality is a rationale for inaction.

What explains the variability in risk of regulated versus unregulated substances?

Factors other than exposure, hazard, and population size also must affect the decisions. Technological, economic, and social factors may be inferred to have an impact on the decisions that led to the "postregulation" and "not significant" data in the previous tables.

FLAWS IN THE SIGNIFICANT RISK CONCEPT

Why can't we have simple definitions of significant or de minimis risk? We can, but they will not mean much. The two notions (and the gap between them) depend not just on the probability of the untoward effect but on the consequences of the effect, whether the risk is undertaken voluntarily, which population is at risk, and so forth. In other words, these notions are tied to individual reactions. They are behaviorally based.

For any individual, risk evaluation is a complicated process. Slovic and his co-workers have empirically analyzed the risk attributes by which laypeople evaluate risk, using psychophysical scaling methods and multivariate analysis.[35] Three groups of attributes (described as dread, familiarity, and number of persons exposed) show a high correlation within group and low correlation between groups, and a correlation with whether a risk is perceived as acceptable. Of these attributes, only the exposed population size appears in our calculations.

Have the courts and Congress played a trick on the public? Perhaps they have. Even in the decision implementing a significant risk policy for OSHA, the Supreme Court noted that it was not intended to be a "mathematical straight jacket." We note six aspects of the other dimensions to significant risk below:

1. The techniques of risk estimation are crude. In any estimate of risk, uncertainties exist that are difficult to evaluate quantitatively. It is exactly the uncertainty in the risk estimate, however, that is of greatest interest in the wide range around significant and de minimis risk magnitudes, where most regulatory decisions occur. The variance in a best estimate of risk for health effects is high, typically in the 10^3 to 10^6 range, and this has important consequences for the evaluation of control measures. Most assessments made by regulatory agencies are upper bounds, with lower bounds of zero. The implications of this range seldom get addressed.

2. From a technical perspective, a number of factors other than the expected magnitude of risk will enter into risk characterization. While Slovic and co-workers have shown that these factors can collapse into three attributes of laypeople, it is not clear that their perceptions will improve a technical analysis (as opposed to the reception of a technical analysis by the public). In this regard, Litai has provided a theoretical list.[36] From a study of a broader range of risk decisions than we have considered here, he came up with eight dichotomous factors that group all public risks into 256 categories. The factors were (1) delayed/immediate risks, (2) necessary/luxury exposure, (3) ordinary/catastrophic risks, (4) controllable/uncontrollable risks, (5) voluntary/involuntary risks, (6) natural/man-made risks, (7) occasional/continuous exposures, and (8) old/new risks. To his list, we would add that frequency estimates often will distort perception of a risk.

    A risk will be viewed differently, for example, if it is stated as a loss of life expectancy. In addition, when several kinds of health effects are at stake, the severity measures are often incommensurate and incoherent.

3. Other modes of societal control besides government regulation come into play, such as insurance, tort liability, and voluntary standards.

4. The technology-driven standards that can come into play after a *significant risk* trigger initiates action are inherently inefficient, because the state of the art in technology does not advance directly in response to environmental needs. Different risk magnitudes may be implemented, just because of the control technology available at the time. As we described in the introduction, standards driven by economic feasibility similarly lead to inefficiencies.

5. As we have demonstrated in the tables above, what we can learn from past decisions is quite limited. Rough guidance is possible, based on precedents, but each significant risk decision is affected by many factors, and the decisions are not transparent. The effects of regulatory decisions are intrinsically difficult to verify. Under our definition of significant risk, once the risk is reduced below these magnitudes, measurement of outcomes is not possible. If it were possible, the risk would still be significant.

6. Ethical considerations influence decisions about risks.[37] Whether a significant risk decision will be accepted will depend on an individual's ethical beliefs. The legal framework in the United States under which regulations are issued has intrinsic ethical values, such as "equal access to the law," which can lead to an egalitarian attitude toward the distribution of risk.

    Clearly, if agencies have little ability to predict what risks an individual will find significant, the Supreme Court's policy is not of much help. Even worse, if there is wide divergence across individuals about what is a significant risk in a given situation, the agencies cannot make point decisions that will be broadly accepted. The traditional techniques of consensus building seem doomed from the start.

    Unfortunately, what appeared to be a wonderful concept has turned out not to be a foundation on which to build regulatory decisions. Therefore, we suggest that simple definitions of significant risk are useful only in a simple world that does not relate well to current realities. If the constraints of the three statutes are accepted, then a significant risk concept can be useful for rough guidance, internal consistency, and relative efficiency. Nothing in the concept removes the overall inefficiency, however, that is inherent in standards based on technology and economic feasibility.

DECISION MAKING IN RISK REGULATION

    The most difficult aspect of risk regulation is setting goals. The Court's notion of a significant risk was meant to finesse a goal-setting problem by getting agreement on what was thought to be an easier notion. Unfortunately, that policy simply hasn't worked so far. We propose a different approach to avoiding the goal-setting problem, namely, increasing efficiency. The basic idea is to have each agency prioritize risks, starting with the one that saves lives at the lowest cost per life. As an agency progressed down its list, it would get to risks that were more and more expensive per life saved. At some point there would have to be a judgment that enough was enough and that further risk reduction was unwarranted. However, that decision is some years away and might never be required, in view of the rate at which new risk situations are discovered. Meanwhile, our proposal is without a de minimis level. Very low and hypothetical risks could be regulated, if the cost-effectiveness turns out to be highly favorable.

The basis of this proposal is an analysis of the effectiveness of each risk reduction in terms of the number of lives saved and of the dollar cost. All proposals would then be ranked in terms of their "cost-effectiveness"—the number of dollars required to save each life. Without a need for slavish adherence to the ordering, an agency would work its way down the list starting with the most cost-effective proposals. The contrast with the current approach of listing because of hazard is fairly obvious.

Our proposal would use society's limited resources as effectively as possible; for a limited amount of resources spent for risk reduction, this proposal would gain the greatest possible risk reduction. If some constituency group wanted to move their proposal to the top of the list, they would have to show why it was more deserving than proposals that were more cost-effective. With such a list, there would be a natural set of advocates against special placement of some proposal, since it would mean putting off the time when other proposals could be pursued. At the moment there is no natural opponent to immediate action at Love Canal, since there is limitless governmental money or at least it is unclear who the money is being taken from. Under the new proposal, the opportunity cost of stringent action at Love Canal would be clear.

To be sure, there are problems with this proposal. It focuses on total social resources needed to lower risks, whereas people tend to think differently about the government's expenditures than they do about the expenditures by the company causing the risk. The cost-effectiveness index above considered only cancer deaths. There are many other risky outcomes. Some way must be found to factor them in. It is not irrational to use cancer primarily, since it usually is expected that the lowest exposure level would be required to protect against carcinogenic outcomes, but this is not always the case. People react differently to risks to various groups, to the precise way the harm is manifested, to the kind of hazard, to the number of people that might be harmed, and to the probability of harm. Each of these dimensions might be required to add structure in addition to the simple cost-effectiveness ranking.

The data on control feasibility and cost often are unavailable or difficult to obtain, and benefits difficult to estimate. For such cases, agencies must have a guideline for which cases to investigate. We suggest the product of the probability of death and the population at risk as a rough way of setting priorities.

Economists insist that it is total social resources that count rather than the cost to any individual in computing efficiency. If Congress desires, it can insist that those who create a risk situation pay to fix it. Whether Congress insists, and whether it is possible, ought not to affect what is done and when it is done. It makes no sense to spend private funds liberally to restore one situation while stinting in another, because the government must bear the cost (or vice versa). If there is a desire to punish a perpetrator, this can be done via fines. Large, well-publicized fines seem an excellent way to obtain prevention, as well.

Most risk situations can result in outcomes ranging from slight bumps to permanent disability to immediate death. The public feared polio because of the disability and seemed to regard the disability as worse than death. An analysis that ignored polio because it focused only on death would be severely deficient. Actually, this problem is the same one that agencies currently face in making decisions; the principal difference is that some explicit basis for comparison would be required.

The National Highway Transportation Safety Administration (NHTSA) faced this difficulty in its 1977 Highway Needs Study, which ranked hundreds of proposals for lowering the injury rate on highways.[38] NHTSA was able to come up with a cost-effectiveness ratio for each of the proposals and to rank them. That study came to some surprising conclusions as to where attention ought to be focused. Another application is the statement by the NRC that exposure to ionizing radiation ought to be lowered as long as it costs no more than $1000 per man-rem.[39] This is a cost-effectiveness criterion that implicitly considers both genetic and somatic damage (the damage to the gene pool and the risk of cancer). It has proved a helpful guideline, and the only controversy has centered on whether the amount is too high or too low.

Idiosyncratic reactions to different situations are built into the current regulatory system. Congress enacted different statutory language for each situation and entrusted each area to a different regulatory agency. Our proposal would have each agency set up its own cost-effectiveness index and work from the top. (Note that EPA is essentially a collection of agencies.) Inevitably, some agencies would find that they could save lives much more cheaply than other agencies. We would not tamper with that process but rather would leave it to the Office of Management and Budget and Congress to react to the cost-effectiveness estimates of each agency and decide whether to reallocate responsibility and budget, or to revise the statutory language. The information on which judgments are based would then be made available to the public, perhaps through modification of an existing mechanism, such as a periodic regulatory agenda.

The proposal to regulate by cost-effectiveness does create the possibility of paralysis by analysis. If an agency had to complete a cost-effectiveness analysis for every possible proposal, it might never finish and never be able to take action. Two notions are needed here.

The first notion is that some groups at risk are too small to be a focus of agency attention. For example, if there were an occupational group of 10 or fewer or a population group of 100 or fewer, these populations would simply be too small to warrant agency attention. One could modify this slightly to say that the agency would act if someone else gathered the data and did the analysis, but that the agency would not initiate data collection or analysis. We propose that an occupational group greater than 100, or a general population group greater than 500, would merit consideration of agency data collection and analysis.

The second notion is that of "agency best effort." An agency should begin with a general consideration of which situations are likely to rank near the top of the cost-effectiveness list. Those situations ought to be the focus of analysis. Other situations would be ignored, at least initially. Inevitably, there would be disagreements as to which items should be on the list. The notion would be that outside groups could attempt to persuade an agency by their own data-gathering or analysis effort. This process would tend to help an agency by supplementing its resources, rather than divert it by stopping actions until proposals had been analyzed.

Perhaps it would be best to combine a cost-effectiveness approach with an escape clause for significant risk situations. Above a high risk level, an agency would regulate with a lesser regard for costs, while taking care to balance any "risk–risk" possibilities. (Risk–risk comparisons balance loss of life from one regulatory option against loss of life from another. For example, an agency would examine the risks of replacements.) Certainly, this combination is more appealing than a combination of cost-effectiveness with a de minimis risk policy. A top-down approach is more efficient than a bottom-up approach.[40]

CONCLUSION

The term "significant risk" has been used explicitly in environmental statutes. Agencies have interpreted the meaning. Litigation over the meaning of significant risk also applies directly to the contextual interpretation of these congressional mandates. Thus, the definition has been articulated as a magnitude of risk that leads to observable changes.

Consistent with this interpretation, we suggest that, superficially, the estimation of a significant risk might involve consideration of only a few factors. If a hazard function exists that specifies the expected incidence of an effect as a function of any exposure and some exposure is expected, then we suggest that analysis of projected incidence plus population provides an approximation of risk magnitude. Thus, the determination of a significant risk is potentially amenable to quantitative evaluation.

We also suggest that for a given substance, there is a region of nontrivial but insignificant (NTI) risk between de minimis and significant risks. NTI risks range from approximately 10^{-3} to 10^{-7} individual lifetime risk, or potentially from 10,000 to 0.5 persons affected annually (depending on the size of the exposed population). The NTI region corresponds to the risks for which there is the highest uncertainty of regulatory agency action (rather than uncertainty in the risk itself, which increases as risk decreases).

We have praised the Supreme Court for stopping misguided efforts to lower risks to the limit of technical and economical feasibility. We see a policy of significant risk, however, as fundamentally flawed. Sharpening our analytical abilities to work with this policy is useful only in the sense of working within the existing system, which can be described as bounded rationality. In the short term, the policy can help regulatory agencies, and we have shown how this can be accomplished. In the longer term, we propose an alternative in the form of cost-effectiveness analysis for ranking the risk management proposals within each agency and placing a priority on the most cost-effective. This policy would save the most lives given the expenditure of some amount of societal cost.

We have explored some of the difficulties in implementing our proposal. It could, however, provide help to the regulatory agencies. Furthermore, it could set in motion a dynamic process to improve the quality of assessments (from data gathering to analysis) and of agency decisions, so that risk management would improve over time.

ACKNOWLEDGMENTS. We thank Laura Byrd, Richard Cothern, Walter Gawlak, Richard Guimond, Elizabeth Margosches, Joseph Merenda, Barbara Morrison, David Patrick, Paul Slovic and Harry Torno for their helpful comments on drafts of this chapter.

REFERENCES

1. Occupational Safety and Health Administration, Identification, Classification and Regulation of Potential Carcinogens. *Federal Register 45*: 5001–5296 (1980).
2. Industrial Union Department v. American Petroleum Institute; 448 *U.S.C.* 607 (1980).
3. American Textile Manufacturers Institute, Inc., et al. v. Donovan, Secretary of Labor, et al. 452 *U.S.C.* 490 (1981).
4. D. Kessler, Food Safety: Revising the Statute. *Science 223*: 1034–1040 (1984).
5. Food and Drug Adiministration Policy for Regulating Carcinogenic Chemicals in Food and Color Additives (Proposed) *Federal Register 47*: 14464 (1982).
6. Environmental Protection Agency, *Federal Register 49*: 18, 130 (1984).
7. Nuclear Regulatory Commission, Safety Goals for Nuclear Power Plants: A Discussion Paper (NUREG 0880; February, 1982); Revision I: Safety Goals for Nuclear Power Plants (May, 1983).
8. Statutes using *significant risk:*

   a. Uranium Mill Tailings Radiation Control Act of 1978, 2, 42 U.S.C. 7901 (1982).
   b. Toxic Substances Control Act, 2, 15 U.S.C. 2601 (1982).
   c. Naturel Gas Policy Act of 1978, 2, 15 U.S.C. 3301 (1982).
   d. Solid Waste Disposal Act Amendments of 1980, 1, 42 U.S.C. 6901 (1982).
   e. Intervention on the High Seas Act, 2, 33 U.S.C. 1471 (1982).
   f. Deepwater Port Act of 1974, 2, 33 U.S.C. 1501 (1982).

9. Regulations using *significant risk*:

   a. Drug Programs, 28: *CFR* 550 (6/29/79; DOJ).
   b. Domestic Quarantine Notices, 7: *CFR* 301 (12/16/67; Animal and Plant Health Inspection Service, USDA).
   c. Domestic Licensing of Production and Utilization Facilities, 10: *CFR* 50 (1/19/56, 3/3/75) (NRC).
   d. Procedures for Compliance with the National Environmental Policy Act, 12: *CFR* 408 (8/30/79) (CEQ).
   e. Assessing the Environmental Effects Abroad of EPA Actions, 40: *CFR* 6.1001 (1/14/81) (EPA).
   f. Pilot Rules (DOT), 33: *CFR* (4/15/82) (CGD).
   g. General Specifications and General Restrictions for Provisional Color Additives for Use in Foods, Drugs, and Cosmetics, 21: *CFR* 81 (3/22/77) (FDA).
   h. Investigational Device Exemptions (FDA), 21: *CFR* 812 (1/18/80) (FDA).
   i. Banned Devices, 21: *CFR* 895 (5/18/79) (FDA).
   j. Administrative Functions, Practices, and Procedures, 21: *CFR* 1316 (4/24/71; 9/24/73) (FDA).
   k. Geological and Geophysical Explorations on the Outer Continental Shelf, 30: *CFR* 251 (1/25/80).
   l. National Interim Primary Drinking Water Regulations, 40: *CFR* 141 (12/24/76) (EPA).
   m. General Statements of Policy or Interpretations (CPSC), 16: *CFR* 1009 (11/1/76).
   n. Substantial Product Hazard Reports (CPSC), 16: *CFR* 1115 (8/7/78).
   o. Regulation of Products Subject to Other Acts Under The Consumer Product Safety Act (CPSC), 16: *CFR* 1145 (9/1/77).
   p. Ban of Artificial Emberizing Materials (Ash and Embers) Containing Respirable Free-Form Asbestos (CPSC), 16: *CFR* 1305 (12/15/77).
   q. Environmental Quality, 14: *CFR* 1216 (1/4/79; NASA).
   r. Protection of Human Subjects (HHS), 45: *CFR* 46 (3/13/75).
   s. General Provisions (DOT), 46: *CFR* 161 (7/10/74).
   t. Vessel Traffic Management, 33: *CFR* 161 (7/10/74; DOT).
   u. Navigational Safety Regulations (DOT), 33: *CFR* 164 (1/31/77).

v. Facilities Engineering, Natural Resources—Land Forest and Wildlife Management (USA), 32: *CFR* 642 (03/28/77).

w. Environmental Protection and Enhancement (USA), 32: *CFR* 650 (12/29/77).

x. Reporting of Defects and Noncompliance (NRC), 10: *CFR* 21 (06/6/77).

y. Patent Licensing Regulations (DOE), 10: *CFR* 781 (11/4/80).

10. The Occupational Safety and Health Act of 1970, Pub. L. 91-596, December 29, 1970, 84 Stat. 1590.

11. The Clean Air Act Amendments of 1977, Pub. L. 95-95, August 7, 1977, 91 Stat. 685.

12. Ethyl Corp, V. EPA; 541 F.2d 1, 26 (D.C. Cir., 1976).

13. Secretary of Labor, Mine Safety and Health Administration (MSHA) v. Cement Control Division, National Gypsum Co. FMSHRC (Federal Mine Safety and Health Review Commission) Docket No. VINC 79-154-PM, April 7, 1981.

14. U.S. Department of Health, Education and Welfare, U.S. Decennial Life Tables for 1969–1971. DHEW Publication No. (HRA) 75-150; National Center for Health Statistics; Rockville, Md. (1975).

15. B. L. Cohen and I. S. Lee, A Catalogue of Risks, *Health Physics 36*: 707–722 (1979).

16. A. S. Whittemore, Quantitative Theories of Oncogenesis. *Advances in Cancer Research 27*: 55–88 (1978).

17. J. C. Rock, The NIOSH Action Level—A Closer Look.Chapter 29 in: *Chemical Hazards in the Workplace*, G. Choudhary (Ed.), ACS Symposium Series 149; Washington, D.C. (1981).

18. National Research Council; Committee on the Biological Effects of Ionizing Radiation, The Effects on Populations of Exposure to Low Levels of Ionizing Radiation. National Academy Press; Washington, DC (1980).

19. P. Milvy, A General Guideline for Management of Risk from Carcinogens. *Risk Analysis 6*: 69–79 (1986).

20. Occupational Safety and Health Administration, Occupational Exposure to Inorganic Arsenic: Supplemental Statement of Reasons for Final Rule. *Federal Register 48*: 1864 (1983).

21. Occupational Safety and Health Administration, Occupational Exposure to Ethylene Dibromide; Notice of Proposed Rulemaking. *Federal Register 48*: 45956–46003 (1983).

22. Occupational Safety and Health Administration, Occupational Exposure to Ethylene Oxide; Final Standard, *Federal Register 49*: 25734–25809 (1984).

23. Occupational Safety and Health Administration, Occupational Exposure to Asbestos; Proposed Rule and Notice of Hearing. *Federal Register 49*: 14116–14145 (1984).

24. Environmental Protection Agency, National Emissions Standards for Hazardous Air Pollutants; Addition of Coke Oven Emissions to List of Hazardous Air Pollutants. *Federal Register 49*: 36560–36564 (1984).

25. Environmental Protection Agency, 4,4'-Methylenedianiline; Initiation of Regulatory Action. *Federal Register 48*: 42898–42900 (1983).

26. Environmental Protection Agency, Toxic Substances; 1,3-Butadiene; Initiation of Regulatory Action. *Federal Register 49*: 20524–20528 (1984).

27. Environmental Protection Agency, National Emission Standards for Hazardous Air Pollutants; Background Information Document (Integrated Risk Assessment) Final Rules for Radionuclides. Office of Radiation Programs; October 22, 1984; EPA 520/1-84-022-2.

28. Environmental Protection Agency, National Emission Standards for Hazardous Air Pollutants; Proposed Standards for Benzene Emissions from Coke By-product Recovery Plants; Proposed Rule and Notice of Public Hearing. *Federal Register 49*: 23522–23555 (1984).

29. Environmental Protection Agency, National Emission Standards for Hazardous Air Pollutants; Benzene Equipment Leaks (Fugitive Emission Sources); Final Rule. *Federal Register 49*: 23498–23520 (1984).

30. Environmental Protection Agency, National Emissions Standards for Hazardous Air Pollutants; Proposed Standards for Inorganic Arsenic. *Federal Register 48*: 33112– 33180 (1983).

31. Environmental Protection Agency, Formaldehyde: Determination of Significant Risk; Advance Notice of Rulemaking and Notice. *Federal Register 49*: 21870– 21897 (1984).

32. Occupational Safety and Health Administration, Coke Oven Emissions, 29: *CFR* 1910.1029 (10/22/76; 1/18/77; 5/23/80).

33. Environmental Protection Agency, Benzene Emissions from Maleic Anhydride Plants, Ethylbenzene Styrene Plants, and Benzene Storage Vessels; Proposed Withdrawal of Proposed Standards. *Federal Register 49*: 8386–8391 (1984).

34. Environmental Protection Agency, National Emission Standards for Hazardous Air Pollutants; Vinyl Chloride. *Federal Register 50*: 1182–1201 (1985).

35. P. Slovic, B. Fischhoff and S. Lichtenstein, Behavioral Decision Theory Perspectives on Risk and Safety. *Acta Psychologica 56*: 183–203 (1984).
36. D. Litai, A Risk Comparison Methodology for the Assessment of Acceptable Risk. Ph.D. Thesis; Massachusetts Institute of Technology; Cambridge, Mass. (1980).
37. R. L. Keeney, Ethics, Decision Analysis, and Public Risk. *Risk Analysis 4*: 117–129 (1984).
38. U. S. National Highway Transportation Safety Administration, *Highway Needs Study: 1981 Update of 1976 Report to Congress*, U.S. Department of Transportation, Washington, D.C., DOT-HS-806-283, (1981).
39. Nuclear Regulatory Commission, 10: *CFR 50*, Appendix I, (1977).
40. P. F. Deisler, Jr., A Methodology for Reducing Industrially Related Cancer Risk. In: *Reducing the Carcinogenic Risks in Industry*, Paul F. Deisler, Jr. (Ed.), Marcel Dekker, Inc.; New York (1984).

6

On Defining a De Minimis Risk Level for Carcinogens

Curtis C. Travis and Samantha A. Richter

INTRODUCTION

Several attempts have been made to use the variation in the levels of natural background radiation to define an acceptable level risk for man-made radiation. The philosophical basis for such proposals is that since no correlations have been detected between variations in natural background radiation and adverse health effects, small additions to natural exposure should be acceptable. The difficulty lies in defining "small." In 1978, Adler and Weinberg proposed using the standard deviation of background radiation levels as a method for establishing radiation exposure limits (Adler and Weinberg 1978). The Adler and Weinberg proposal results in the suggestion that a lifetime cancer risk of about 10^{-4} is de minimis.* The Adler and Weinberg de minimis risk level was based on the standard deviation of human exposure to background terrestrial and cosmic radiation. We propose to determine the risk levels associated with the standard deviation of human exposure to other radioactive and chemical carcinogens.

For this study, sufficient data have been collected to characterize the mean and standard deviation of environmental concentrations of three sources of human exposure to naturally occurring radionuclides: external exposure to background radiation, inhalation of background radiation, and ingestion of radionuclides in drinking water. We have also characterized the mean and standard deviation of measured concentrations of three background chemical carcinogens: benzene and formaldehyde in air, and chloroform in drink-

---

* There are several definitions of a de minimis risk, but we will use the term to designate a level of risk that is below regulatory concern.

Curtis C. Travis and Samantha A. Richter • Health and Safety Research Division, Oak Ridge National Laboratory, Oak Ridge, Tennessee, 37830.

ing water. We propose to determine what, if any, variation exists in the standard deviations of concentrations of these carcinogens, and if the de minimis level of risk resulting from these exposures deviates significantly from the 10^{-4} level obtained by the Adler and Weinberg proposal.

APPLICATION OF THE NATURAL BACKGROUND-STANDARD DEVIATION METHOD TO RADIONUCLIDE CONCENTRATIONS

Measurements were obtained on natural background levels of selected radionuclides in order to calculate variations from the mean and associated risk levels. Graphs of the data indicate that concentrations of radionuclides from the three sources are log-normally distributed; therefore, the proper measure of variation from the mean is the geometric standard deviation.

External Exposure to Natural Background Radiation

There are three major sources of natural background radiation in the United States: terrestrial, cosmic, and natural radionuclides deposited in the body. Large variations exist in the levels of terrestrial and cosmic radiation among the 50 states, ranging from 15–35 mrem per year in the Atlantic and Gulf Coast areas to 75–140 mrem per year along the Colorado plateau (NAS/NRC 1980). In 1977, Adler and Weinberg (Adler and Weinberg 1978) estimated the standard deviation of the average natural background terrestrial and cosmic radiation level in the United States to be approximately 20 mrem per year and suggested that this dose be the maximum allowable exposure to an individual in the public from operation of the entire uranium fuel cycle. Analysis of natural background radiation data collected in 1972 (Oakley 1972) and 1981 (which are essentially the 1972 measurements with an 0.64 shielding correction; Bogen and Goldin 1981) indicates that the standard deviation of background radiation levels may not be as high as 20 mrem per year. Figure 1 plots radionuclide levels in each of the 50 states. The mean level using the 1972 Oakley data is 84 mrems per year, with a geometric standard deviation of 1.1, which corresponds to 12 mrem per year. The mean level of the 1981 Bogen/Goldin data is 56 mrem per year, with a geometric standard deviation of 1.2 or 9 mrem per year.

Assuming that an annual whole body exposure to 1 rad low-LET radiation results in a lifetime (70-year) risk of fatal radiogenic cancer of 110 fatalities per million person rad, and setting 1 rad equal to 1 rem (NAS/NRC 1972, 1980), we calculated the individual lifetime cancer risks associated with the above standard deviations (Table 1).

Inhalation of Radon

Radon is emitted from trace concentrations of radium in the earth's crust and from building materials such as rock, brick, and concrete (Evans et al. 1981). Recent studies have measured naturally occurring concentrations of radon daughters in homes, offices, and schools. Whereas the risk from terrestrial and cosmic radiation considered in the previous section results from external exposure, the risk from radon results from inhalation. In 1984, R. A. Oswald published data on indoor radon measurements taken in 31

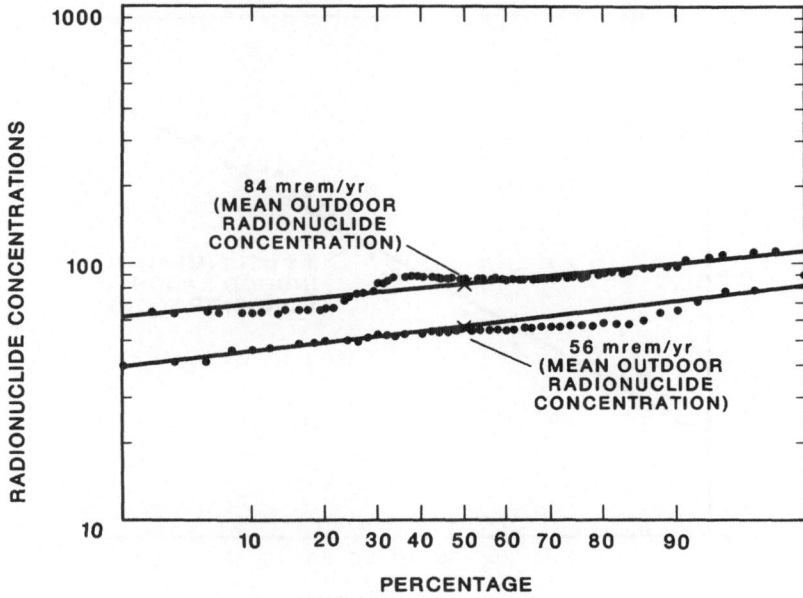

Figure 1. Cumulative distribution of outdoor terrestrial and cosmic radiation concentrations from 50 states. Data analyzed in 1972 indicate a mean concentration of 84 mrems per year. Data analyzed in 1981, which are essentially the 1972 data with an indoor shielding correction, yield a mean level of 56 mrem per year.

states in the United States (Oswald 1984). Using the mean recorded level from each of the 31 states, he found the overall mean to be 1.4 pCi/l and the geometric standard deviation to be 2.4, or 1.9 pCi/l (Figure 2). This compares well with the geometric standard deviation of 2.8, or 2.7 pCi/l for indoor radon reported by Nero following analysis of 19 data sets (Nero 1985). Assuming a lifetime (70-year) radon cancer risk of 1.5×10^{-4} per working level month (WLM), and setting 1.2 WLM equal to 10 pCi/l (NCRP 1984), the individual lifetime cancer risk associated with 1.9 pCi/l is 2.2×10^{-3}. The outdoor/indoor radon activity ratio is about 0.086 (UNSCEAR 1982). Therefore, assuming that outdoor and indoor radon have similar probability distributions, the standard

Table 1. Standard Deviations of Natural Background Terrestrial Cosmic Radiation and Associated Risk Levels

Standard deviation	Lifetime risk
20 mrem/year [a]	1.5×10^{-4}
12 mrem/year [b]	9.2×10^{-5}
9 mrem/year [c]	7.0×10^{-5}

[a] Adler/Weinberg calculation (1978).
[b] Calculated using Oakley data (1972).
[c] Calculated using Bogen/Goldin data (1981).

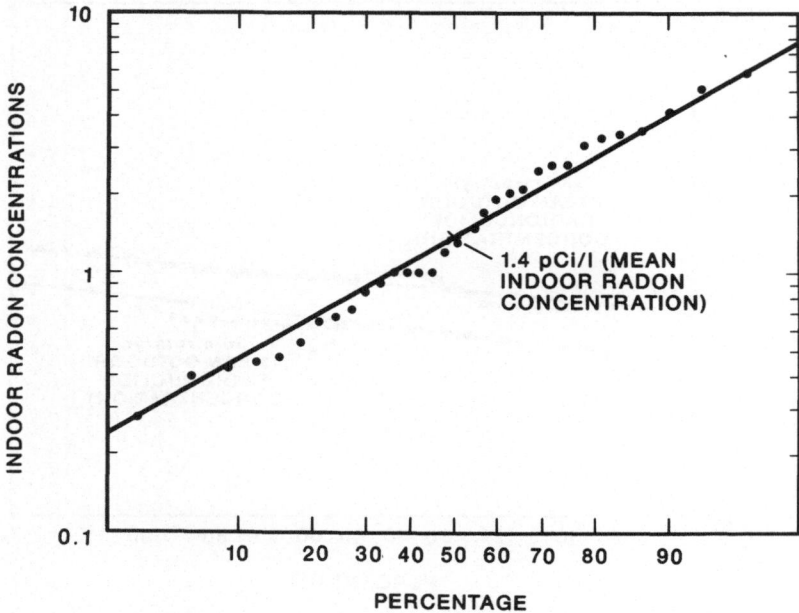

Figure 2. Cumulative distribution of indoor radon concentrations taken in 31 states. The data indicate a mean indoor radon concentration of 1.4 pCi/l.

deviation associated with the mean concentration of outdoor radon is about 0.16 pCi/l with an associated individual lifetime cancer risk of about 1.9×10^{-4}. Since individuals spend 90% of their time indoors (Evans et al. 1981), the total individual lifetime cancer risk associated with the standard deviations of radon concentrations is about 2.4×10^{-3}.

Radionuclides in Drinking Water

Radium-226 and Radium-228.   Radium-226 and radium-228 are long-lived daughters of the uranium-238 decay series and occur as trace constituents in granites and metamorphic rocks (Michel and Moore 1980). They are the most hazardous of the naturally occurring radionuclides in drinking water (EPA 1975). Measurements of radium-226 in 50,000 public drinking water supplies nationwide (Watson, Etnier, and McDowell-Boyer 1983) range from undetectable in Brookfield, Connecticut, to 196 pCi/l in Alamo, Georgia. Figure 3 shows the mean radium-226 drinking water concentration to be 0.58 pCi/l. This compares well with a reported population-weighted average range of between 0.3 and 0.8 pCi/l (Cothern 1985). The geometric standard deviation associated with the mean is 2.1, or 0.62 pCi/l. Assuming that the 1 pCi/l of radium- 226 in drinking water results in an annual bone dose of 30 mrem and a lifetime risk of approximately 2.5×10^{-5} (EPA 1975), the individual lifetime cancer risk associated with the standard deviation of radium-226 concentrations in drinking water supplies is 1.5×10^{-5}. The radium- 228/radium-226 activity ratio in natural water is about 1.2 (Cothern, Lappenbusch, and Michel 1986)

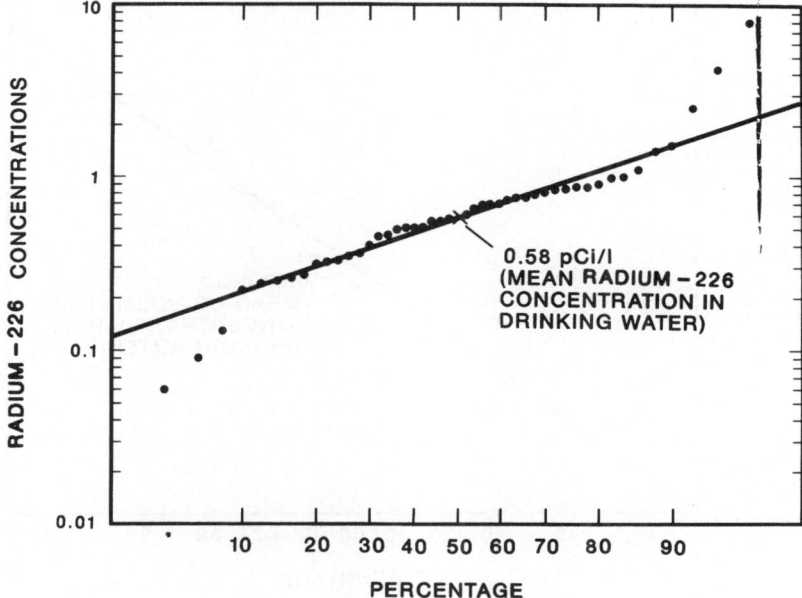

Figure 3. Cumulative distribution of radium-226 concentrations in drinking water taken from public water supplies in 50 states. The data indicate a mean radium-226 concentration in drinking water of 0.58 pCi/l.

and 1 pCi/l of radium-228 in drinking water results in a lifetime cancer risk of 1.6×10^{-5} (Cothern et al. 1986). Therefore, the standard deviation associated with the mean radium-228 concentration in drinking water is about 0.74 pCi/l, which results in an individual lifetime cancer risk of about 1.1×10^{-5}.

Uranium-238. Uranium occurs naturally in geological formations, in mineral deposits, and in sea water, all of which may contact possible drinking water sources (CRPB 1980; Turekian and Chan 1971). The uranium-238 isotope, which makes up 99.27% of natural uranium, poses the greatest risk from uranium in drinking water. Measurements of uranium concentrations in 28,239 domestic drinking water supplies (Drury et al. 1981) ranged from 0.07 to 652 pCi/l, with the highest concentrations found in South Dakota. Figure 4 shows the mean uranium concentration to be 0.27 pCi/l. This mean falls close to the population-weighted median concentration range for uranium-238 of 0.1–0.2 reported by Cothern and Lappenbusch (1983). The geometric standard deviation associated with this concentration is 4.8, or 1.03 pCi/l. The lifetime cancer risk of 1.03 pCi/l of uranium in drinking water is approximately 3×10^{-6} (Cothern et al. 1983).

Radium-226, radium-228, and uranium-238 contribute the majority (greater than 90%) of the risk from naturally occurring radionuclides in drinking water. Thus, the individual lifetime cancer risk associated with the standard deviations of radionuclides in drinking water is 2.9×10^{-5}.

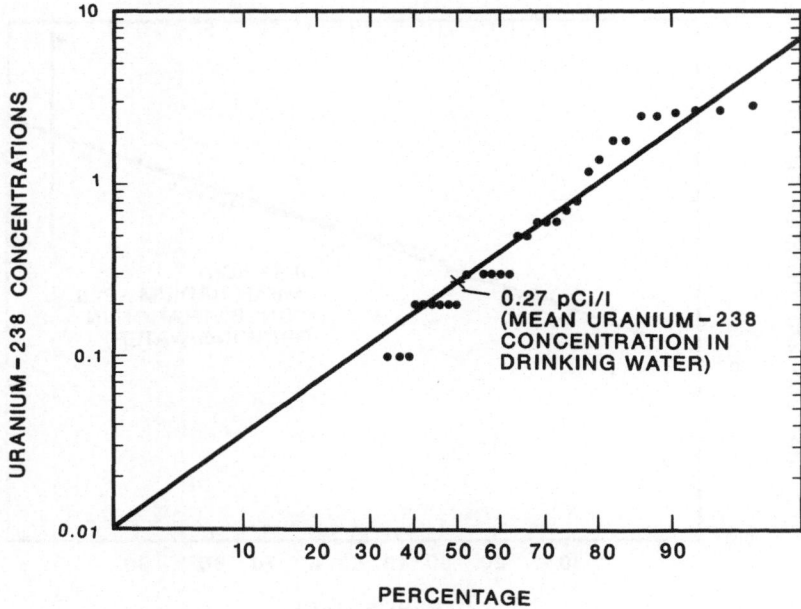

Figure 4. Cumulative distribution of mean uranium-238 concentrations in drinking water taken from domestic drinking water supplies. The data indicate a mean uranium-238 concentration in drinking water of 0.27 pCi/l.

APPLICATION OF THE BACKGROUND-STANDARD DEVIATION METHOD TO CHEMICAL CARCINOGENS

We have also collected sufficient data to characterize the means and standard deviations of environmental concentrations of three ubiquitous chemical carcinogens: benzene and formaldehyde in the air, and chloroform in drinking water. While all three of these chemical carcinogens occur naturally, human activities are responsible for increased human exposures. Thus, the situation is not completely analogous to that of naturally occurring radionuclides. Nevertheless, the risk levels associated with the standard deviation of the enhanced environmental concentrations of these carcinogens can be used to provide upper-bound estimates for a de minimis risk level (if one accepts the Adler and Weinberg proposal that the standard deviation is an appropriate measure of de minimis). Again, since the environmental concentrations of these chemical carcinogens are log-normally distributed, the proper measure of variation from the mean is the geometric standard deviation.

Outdoor Benzene

Benzene is a hydrocarbon naturally occurring in crude oil and emitted to the environment mainly from gasoline evaporation. Thirty-five ambient benzene measurements collected in 1983–1984 in four California cities (ARB-DHS 1984) were plotted to find the mean and geometric standard deviation for background benzene levels (see Figure

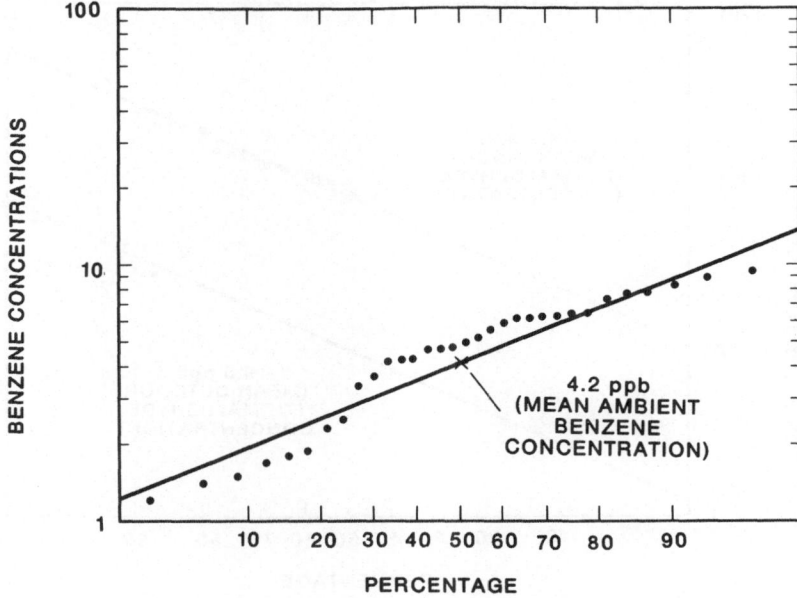

Figure 5. Cumulative distribution of outdoor benzene concentrations taken in four California cities. The data indicate the mean outdoor benzene concentration in the cities is 4.2 ppb.

5). The range of the measurements was very small, from 1.2 ppb in El Monte to 11 ppb in downtown Los Angeles. The mean ambient benzene level was found to be 4.2 ppb, while the geometric standard deviation was found to be 1.7, or 3.4 ppb. Using carcinogen potency data published by the EPA (CAG 1984), the individual upper-bound cancer risk associated with 4.3 ppb benzene is 2.2×10^{-4}.

Indoor and Outdoor Formaldehyde

The major sources of background indoor formaldehyde are formaldehyde resins found in structural materials, insulation, and furnishings. Measurements collected in 40 East Tennessee homes indicate formaldehyde levels as high as 0.4 ppm in a new, well-insulated home in the summer, with the average at 0.04 ppm for new homes and 0.08 ppm for homes older than five years (Hawthorn et al. 1984) (see Figure 6). Several of the formaldehyde concentrations were below detection limits since the measurement devices used for the study could not detect levels below 25 parts per billion. The mean indoor formaldehyde concentration was found to be 74.5 ppb, while the geometric standard deviation was found to be 1.6, or 125.5 ppb. Using formaldehyde potency data published in the *Federal Register* (EPA 1984), the upper bound for individual cancer risk associated with 125.5 ppb is 1.1×10^{-4}.

In contrast to indoor formaldehyde, outdoor formaldehyde levels were always less than 0.03 ppm. Many outdoor concentrations were below detection. Thus, the relative percentage of measurements between 0 and 25 ppb was high (just below 88%). The mean

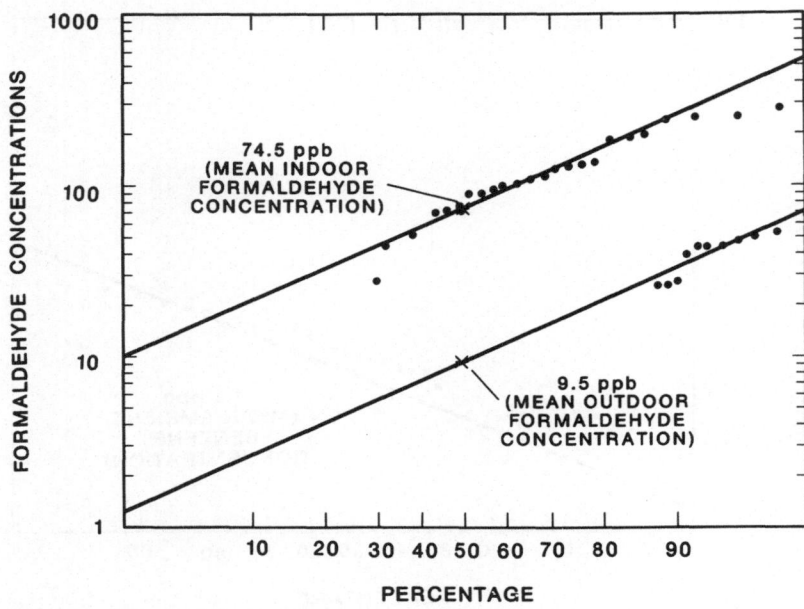

Figure 6. Cumulative distribution of formaldehyde concentrations taken indoors and outdoors in the East Tennessee area. The data indicate the mean indoor formaldehyde concentration in the area to be 74.5 ppb and the mean outdoor concentration to be 9.5 ppb.

outdoor formaldehyde level was found to be approximately 9.5 ppb, while the geometric standard deviation was found to be 1.3, or 12.5 ppb, one order of magnitude less than the standard deviation for indoor formaldehyde levels. The individual cancer risk upper bound associated with 12.5 ppb is 1.1×10^{-5}.

Figure 6 demonstrates one of the difficulties in using the standard deviation to define a de minimis risk level. The two lines in Figure 6 are parallel, indicating that the log-normal distribution for outdoor and indoor formaldehyde concentrations have the same geometric standard deviations. However, since the means of the two distributions differ by a factor of 10, the 68% confidence intervals about the means also differ by a factor of 10. In other words, even though both distributions exhibit the same amount of variability about the mean, the level of risk associated with this variability is a factor of 10 lower for the distribution with the smaller mean (the outdoor case).

Chloroform in Drinking Water

Four major trihalomethanes are found in drinking water: chloroform, bromodichloromethane, dibromochloromethane, and bromoform. The major risk from trihalomethanes in drinking water comes from chloroform. Most chloroform in drinking water results from the water chlorination process, though concentrations have been measured in raw water used for drinking water. Sixty-eight measurements of chloroform concentrations in drinking water supplies (Symons et al. 1975) were plotted to find the mean and geometric standard deviation of chloroform in finished drinking water (see Figure 7). The mean

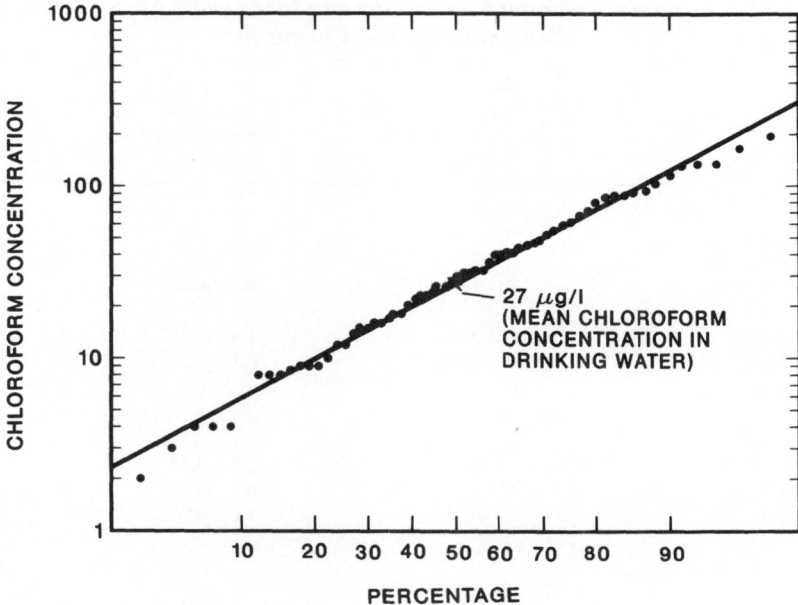

Figure 7. Cumulative distribution of chloroform concentrations in 68 treated water supplies in the United States. The data indicate a mean chloroform concentration in finished drinking water of 27 μg/l.

chloroform level was found to be 27 μg/l, while the standard deviation was found to be 3.3, or 63 μg/l. Assuming that 0.10 mg/l of chloroform in drinking water produces an individual lifetime risk of 1×10^{-4} (EPA 1976), the individual lifetime cancer risk associated with the standard deviation of chloroform in drinking water is 2.2×10^{-4}.

DISCUSSION

Adler and Weinberg proposed using the standard deviation of natural background radiation as a method for establishing a de minimis risk level. We have estimated the variation in natural background concentrations for three exposures to naturally occurring radionuclides and three chemical carcinogens in order to determine the reliability of the Adler/Weinberg technique. Since the environmental concentrations of all of these substances are log-normally distributed, the proper measure of variation about the mean is the geometric standard deviation. Analysis of the risk levels associated with the standard deviations of human exposures to naturally occurring radionuclides indicates that except for exposures to indoor radon, the sum of the risk level associated with naturally occurring radionuclides is about 2.5×10^{-4} (see Table 2). Thus, if one accepts the Alder and Weinberg assumption that the standard deviation of background radiation is the appropriate measure of smallness, it appears that 10^{-4} may be an upper-bound estimate for de minimis for chemical carcinogens.

In recent literature, the Nuclear Regulatory Commission (NRC) has implicitly defined

Table 2. Geometric Standard Deviations and Risk Levels Associated with
Radionuclides and Chemicals

	Geometric standard deviation (GSD)	Concentration corresponding to GSD	Upper bound risk level
Radionuclides			
External exposure			
Background radiation			
Bogen/Goldin data	1.2	9.0 mrem/yr	7.0×10^{-5}
Inhalation			
Indoor radon	2.4	1.9 pCi/l	2.2×10^{-3}
Outdoor radon	2.4	0.16 pCi/l	1.9×10^{-4}
Total radon			2.2×10^{-3}
Ingestion			
Drinking water			
Radium-226	2.1	0.62 pCi/l	1.5×10^{-5}
Radium-238		0.74 pCi/l	1.1×10^{-5}
Uranium-238	4.8	1.03 pCi/l	0.3×10^{-5}
Drinking water			
Total risk			2.9×10^{-5}
Chemicals			
Inhalation			
Benzene			
Outdoor	1.7	3.4 ppb	2.2×10^{-4}
Formaldehyde			
Indoor	1.6	125.5 ppb	1.1×10^{-4}
Outdoor	1.3	12.5 ppb	1.1×10^{-5}
Ingestion			
Drinking water			
Chloroform	3.3	63.0 μg/l	2.2×10^{-4}

several methods for establishing safety levels that could be interpreted as de minimis, three of which are: (1) 1 mrem per year (NRC 1985), (2) 0.1% of U.S. prompt fatalities per year (NRC 1983), and (3) 0.1% of U.S. latent cancer deaths per year (NRC 1983). The lifetime risk associated with 1 mrem per year radiation exposure is 7.7×10^{-6}. The lifetime risk level associated with 0.1% of prompt fatalities is 2.8×10^{-5}, while the estimated lifetime risk associated with 0.1% of U.S. latent cancer fatalities is 1.3×10^{-4}. Thus, the NRC has proposed de minimis risk levels for radionuclides on the order of 10^{-5} or 10^{-4}. These are consistent with de minimis risk levels obtained using the Adler and Weinberg approach.

We now briefly provide another perspective on the problem of defining a de minimis level of risk. Travis et al. (1987) retrospectively reviewed cancer risk data used to support regulatory decisions concerning carcinogenic substances. The sources of the data reviewed were proposed and final regulation notices found in the *Federal Register* and published and unpublished regulatory support documents (all of which are in the public domain). The purpose of the review was to determine a level of risk below which regulatory action historically has not been taken to reduce risk. The measures of risk most often encountered during the review were (1) individual lifetime risk, (2) the size of the population exposed to the risk, and (3) the expected number of annual cancer deaths in the exposed population, referred to as the population risk.

The analysis revealed that regulatory practices cannot be explained on the basis of lifetime risk alone, or the relationships between lifetime risk and population size. The key to understanding regulatory practices is in the relationship between lifetime risk and population risk. The indication from data used by several regulatory agencies is that for small population risks (fewer than 10^{-1} cancer deaths per year in the exposed population), regulatory action was seldom taken on individual lifetime risk levels less than about 1×10^{-4}. As population risk approaches 250 cancer deaths per year (which could occur only in populations the size of the total U.S. population), the de minimis level drops to 10^{-6} (see Figure 8). Taken as a whole, these data indicate that past federal regulatory actions concerning chemical carcinogens have implicitly defined a de minimis level of risk that depends on both individual risk *and* population impact. These data indicate that

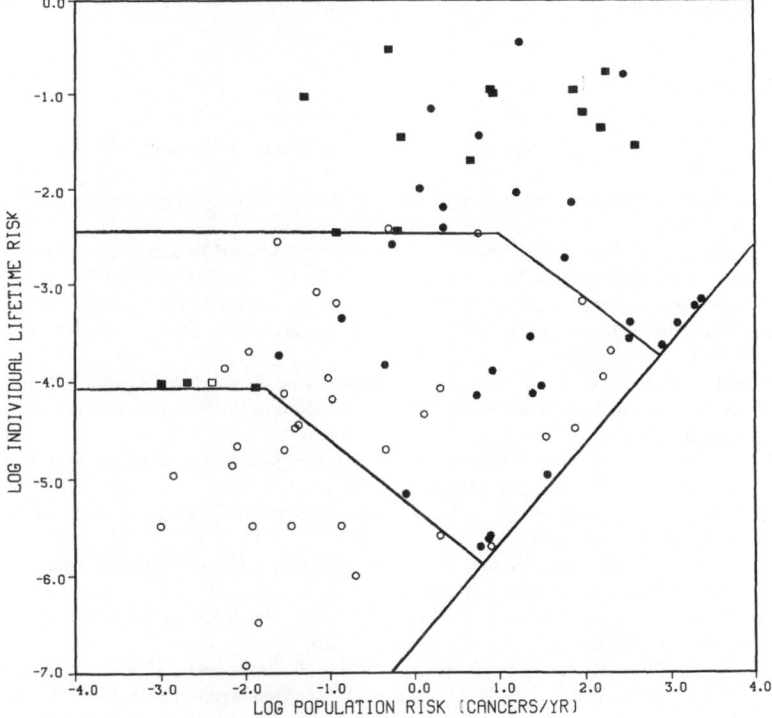

Figure 8. Effect of individual lifetime cancer risk versus population cancer risk (cancer deaths per year) estimates on chemical carcinogen regulation decisions. Open circles indicate that agencies did not act to reduce public risk. Closed circles indicate that agencies did act to reduce public risk. Open and closed squares have the same meaning for occupational risks. The upper line defines a preregulatory level of risk that induces regulation. The lower line defines a preregulatory level of risk that seldom induces regulation. The diagonal line defines the boundary beyond which no data will occur.

for radiation exposures to small populations (fewer than 10^{-1} cancer deaths per year), the Adler and Weinberg proposals of 1×10^{-4} as a de minimis risk level may be appropriate and consistent with past regulatory actions on chemical carcinogens. However, for radiation exposures involving large population impacts, the de minimis level should be reduced to 1×10^{-6}.

CONCLUSION

Analysis of the standard deviation associated with natural background concentrations of radionuclides and chemical carcinogens indicate a lifetime risk on the order of 10^{-4} lower. However, analysis of regulatory practices concerning chemical carcinogens indicates that the de minimis level varies from 10^{-4} to 10^{-6}, depending on the size of the population impact. Since the regulatory process combines many economic, social, and political factors into an implicit expression of society's willingness to accept risk, it is probably superior to the standard deviations of background radiation method for defining de minimis risk levels.

REFERENCES

Adler, H. I. and A. M. Weinberg, "An Approach to Setting Radiation Standards," *Health Physics 34*: 719–720, (June 1978).

Air Resources Board and Department of Human Services, *Report to the Scientific Review Panel on Benzene*, (November 1984).

Bogen, K. T. and A. S. Goldin, *Population Exposure to External Natural Radiation Background in the United States 1960–2000*, Environmental Protection Agency ORP/CSD 72-1 (August 1981).

Carcinogen Assessment Group, "Relative Carcinogenic Potencies among 53 Chemicals Evaluated by the Carcinogen Assessment Group as Suspect Human Carcinogens," U.S. Environmental Protection Agency (October 1984).

Canadian Radiation Protection Bureau, "Section on Uranium from the Canadian Radiation Protection Bureau, Guidelines for Canadian Drinking Water Quality," supporting documentation to Health and Welfare, Ottawa, Canada (1980).

Cothern, R. C., "Techniques for the Assessment of Carcinogenic Risk Due to Drinking Water Contaminants," *CRC Critical Reviews in Press*: 32, (October 17, 1985).

Cothern, R. C., and W. L. Lappenbusch, "Occurrence of Uranium in Drinking Water in the U.S.," *Health Physics 45*(1): 89–99, (July 1983).

Cothern, R. C., W. L. Lappenbusch and J. A. Cotruvo, "Health Effect Guidance for Uranium in Drinking Water," *Health Physics 44*(1): 377–384, (1983).

Cothern, R. C., W. L. Lappenbusch and J. Michel, "Drinking Water Contribution to Natural Background Radiation," *Health Physics 50*(1): 33–47, (January 1986).

Drury, J. S., S. Reynolds, P. T. Owen, R. H. Ross and J. T. Ensminger, "Uranium in the U.S. Surface, Ground, and Domestic Waters," U.S. Environmental Protection Agency, EPA-570/9- 81-001 (1981).

Environmental Protection Agency, "Notice of Proposed Maximum Contaminant Levels of Radioactivity," U.S. Environmental Protection Agency, *Federal Register 40*: 34324, (1975).

Environmental Protection Agency, "Formaldehyde Determination of Significant Risk," *Federal Register 49*: 21870, (May 1984).

Evans, R. D., J. H. Harley, W. Jacobi, A. S. McLean, W. A. Mills and C. G. Stewart, "Estimate of Risk from Environmental Exposure to Radon-222 and its Decay Products," *Nature 290* (March 1981).

Hawthorn, A. R., R. B. Gammage, C. S. Dudney, B. E. Hingerty, D. D. Schuresko, D. C. Parzyck, D. R. Womack, S. A. Morris, R. R. Westley, D. A. White, and J. M. Schrimsher, *An Indoor Air Quality Study of Forty East Tennessee Homes*, Oak Ridge National Laboratory, ORNL-5965 (December 1984).

Michel, J. and W. S. Moore, "^{228}Ra and^{226}Ra content of groundwater in fall line aquifers," *Health Physics* *38*: 336–671, (1980).

National Academy of Sciences, National Research Council, *The Effects on Populations of Exposures to Low Levels of Ionizing Radiation*, BEIR-1 (1972).

National Academy of Sciences, National Research Council, *The Effects on Populations of Exposures to Low Levels of Ionizing Radiation*, BEIR-3 (1980).

National Council on Radiation Protection, *Evaluation of Occupational and Environmental Exposures to Radon and Radon Daughters in the United States*, NCRP Report 78 (1984).

Nero, A. V., "Indoor Concentrations of Radon 222 and Its Daughters: Sources, Range, and Environmental Influences," *Indoor Air and Human Health*, ed. by R. B. Gammage and S. V. Kaye, Lewis Publishers, Chelsea, Mich., 43–67, (1985).

Nuclear Regulatory Commission, *Safety Goals for Nuclear Power Plant Operations*, U. S. Nuclear Regulatory Commission NUREG-0880, Rev. 1 (May 1983).

Nuclear Regulatory Commission, *Proposed Revision of 10 CFR Part 20, "Standard Protection Against Radiation,"* U.S. Nuclear Regulatory Commission SECY-85-147-Part 1 (April 1985).

Oakley, D. T., *Natural Radiation Exposure in the United States*, Office of Radiation Programs, Environmental Protection Agency (1972).

Oswald, R. W., "Indoor Radon Results in the United States," Terradex Corporation (June 1984).

Symons, J. M., A. Thomas, J. Ballor, C. Keith, J. DeMarco, K. K. Kropp, G. G. Robeck, D. R. Seeger, C. J. Slocum, B. L. Smith and A. A. Stevens, "National Organics Reconnaissance Survey for Halogenated Organics," *Journal AWWA*: 634– 652, (November 1975).

Travis, C. C., S. A. Richter, E.A.C. Crouch, R. Wilson, and E. Klema, "Cancer Risk Management", *Environmental Science and Technology* 21:415–420, (1987).

Turekian, K. K. and L. H. Chen, "The Marine Geochemistry of Uranium Isotopes, 230Th and 231Pa, Activation Analysis," *Geochemistry and Cosmochemistry*, ed. by A. O. Brunfelt and E. Steinnes, UNIVERSITET-SPOR-LAGET, Oslo-Berger-Tromso (1971).

United Nations Scientific Committee of the Effects of Atomic Radiation, "Ionizing Radiation: Sources and Biological Effects," 19, U.N., New York (1982).

Watson, A. P., E. L. Etnier and L. M. McDowell-Boyer, *Radium-226 in Drinking Water and Terrestrial Food Chains: A Review of Parameters and an Estimate of Potential Exposure and Dose*, ORNL/TM-8597 (1983).

7

De Minimis Risk and the Integration of Actual and Perceived Risks from Chemical Carcinogens

Paul Milvy

We all face a nontrivial lifetime risk of death equal to 1. It is against this ultimate benchmark that all risks must be evaluated. But we all recognize that we also have a considerable diversity of opinions, of values, of analytical approaches, and of philosophy in terms of how best to handle all the component risks, including those for which society is responsible.

Risk management is composed of both objective and subjective dimensions. Each is equally appropriate and germane to risk management. Although some of the subjective factors are codified in laws, ethics, cultural norms, and philosophical world views and have thereby achieved a veneer of objectivity, they remain subjective, albeit sanctioned by societal consensus. On the other hand, we must acknowledge that much of what we think of as objective and scientific also has subjective aspects.

Risk to life and limb is a quality of the world we live in that is objective and can be objectively studied and evaluated. The rates of morbidity and mortality that we measure and tabulate are records of the slings and arrows (many man-made, many not) to which we are all susceptible. And these statistics, as has been said, are simply the people who are at risk with their tears wiped away. The process of data collection and analysis is largely objective, and when the science is good, a scientific consensus can be expected quickly to emerge. On the other hand, considerable diversity of opinion is expressed when the attempt is made to define a rate of risk that may be regarded as trivial, acceptable,

The views expressed in this chapter are the author's and do not necessarily reflect those of the Environmental Protection Agency.

Paul Milvy • Environmental Protection Agency (WH550 D), Washington, D.C. 20460.

or negligible. The phrases "acceptable risk" and "de minimis risk" have adjectives that are inherently qualitative and subjective in character. They modify a noun that in theory is objectively quantifiable.

Dictionary definitions do not provide guidance; they tend rather to beg the question. But for this very reason, they are instructive. Several dictionaries define "acceptable" only in a self-referential way. Webster's definition is marginally better: It defines "acceptable" as "satisfactory," and "barely satisfactory or adequate." "Trivial," the operative word in the legal definition of de minimis, is defined as "of little worth or importance" and "insignificant." "Trifle" is defined as "an insignificant or relatively small amount." Thus, the terms by which we seek to define de minimis are themselves defined by words equally subjective, qualitative, and imprecise.

We are unable to exclude a subjective dimension to risk management because a subjective—human—dimension constitutes one of the two pillars that support the entire conceptual edifice.

Failure to use a judicious mix of subjective and objective factors in the determination of de minimis risk levels is counterproductive and doomed to failure. Although risk managers, often trained in the physical sciences, statistics, or mathematics, might view reliance upon collective common sense and gut reaction to be a less than ideal approach, the reasons that such subjective dimensions are legitimate components of the task of establishing de minimis risk criteria are not difficult to identify.

If it is accepted that attempts to base risk management judgments wholly on objective criteria will fail, another approach must be sought. It is the goal of this exercise to develop a criterion for de minimis carcinogenic risk that integrates into a single expression both objective and subjective components of risk management.

The subjective dimensions of risk analysis have been objectively studied by many investigators.[1-3] Litai, Lanning, and Rasmussen[4] have, in particular, shown that risks can be categorized using pairs of antonyms (e.g., old/new, voluntary/involuntary, natural/man-made) and that we perceive different categories to be of different concern. Litai quantified the subjective evaluation of these categories, and we can infer that extreme categories (e.g., a delayed, controllable, voluntary, old hazard vs. an immediate, uncontrollable, involuntary, new one) differ by many orders of magnitude on a scale of perceived risk.

The subjective quality of objective approaches that seek to develop an acceptable de minimis risk is also illustrated by two well-known quantitative analyses that sought to quantify a de minimis guideline or cutoff to regulate human exposure from societal radiation sources. In 1978 Adler and Weinberg[5] suggested a de minimis risk of 20 mrem per year that was based upon the standard deviation of a population- weighted annual state-to-state variation in the natural background rate. A second de minimis radiation criterion was suggested by G. H. Whipple in 1980[6] and was based upon the standard deviation of the mean of 20 years of annual U.S. cancer mortality data. The value for the de minimis risk for annual radiation dose calculated by these two approaches differ by a factor of 250. I suspect few experts view this disparity as trivial, even in the absence of a satisfactory definition for this word!

These early and interesting approaches to the estimation of de minimis risk—essentially estimating the size of the random noise—can easily be employed to develop widely disparate values for risk criteria. For example, using annual U.S. cancer mortality statistics from 1968–1982,[7] U.S. total monthly mortality for 1982–1983[8] (normalized for months of equal duration; 30.42 days), and appropriate U.S. population figures, standard deviations can be calculated that vary by three orders of magnitude. Table 1 indicates the results, which vary by up to a factor of a thousand. It is clear that by selecting appropriate data (e.g., U.S. mortality rates from rare tumors) ratios of standard deviations of the mean to the population can be determined that are very much smaller than the lowest rate shown in Table 1. What criteria, then, are to be used to select from these multiple potential de minimis criteria the single criterion appropriate for the risk management of carcinogenic hazards?

In a very real sense (I generalize from my own experience and from trying to probe my own motives when I examine these problems), the analyst gropes for an analytical methodology that results in a level of "trivial" risk that is consistent with the analyst's own subjective value system, which is of course itself to a significant degree determined by the ambient culture. If such an analytical approach is found, it then seems to confer upon the result an imprimatur of value-free objectivity. This scientific veneer tends to hide the subjective origins of the criterion.

The above approach seeks to define a de minimis risk level by determining an exposure that is so small that, although real, it is buried in the random noise of statistical fluctuation. A second approach, historically more common, is to select a small fraction of total risk and to argue that this small incremental mortality is trivial compared to total population mortality.[9] Thus we say, for example, that because the annual risk of death for those of working age, 18 to 65, is 1 in 256, an incremental increase in mortality over a lifetime of 1 in 1 million (10^{-6}), or perhaps 1 in 100,000, is acceptable or trivial. A 10^{-6} lifetime risk is often cited as an appropriate level of acceptable risk for man-made carcinogenic chemicals and is, if not invariably achievable, certainly considered by many to be a level for which regulatory authorities should strive.

The size of the population at risk does not explicitly influence the selection of the level of risk judged to be de minimis or acceptable. Indeed, the argument has been made

Table 1. De Minimis Risk Examples Based upon Standard Deviations of Selected Data

Data	SD of data	SD of mean[d]
Monthly[a] mortality/population × 70 years	2×10^{-3}	5×10^{-4}
Annual[b] cancer death rates	2×10^{-4}	5×10^{-5}
1982 monthly mortality/pop. quadratic eq.[c]	2.2×10^{-5}	6×10^{-6}
1982–1983 monthly mortality/pop. quadratic eq.[c]	2.11×10^{-5}	6×10^{-6}
Annual[b] cancer death rates quadratic eq.[c]	—	2×10^{-6}

[a] All monthly data adjusted to average days/month = 30.42 days. 1982–1983 data.
[b] 1968–1982 data. The standard deviation of the 1982–1983 data is given to three significant figures because the result is used below.
[c] The standard deviations are calculated for the quadratic equation that best fits the data.
[d] This is obtained by dividing the standard deviation of the data by the square root of the number of data points.

that it should not be a factor. The rationale for ignoring the size (or density) of the population at risk when setting standards has been well expressed by Doniger, an attorney for a public interest group:

> Why should the degree of protection that a person is entitled to differ according to how many neighbors he or she has? Why is it all right to expose people in lightly populated areas to higher risks than people in densely populated ones?[10]

Such rhetorical questions seem reasonable and have an emotional ring of truth to them as well. A 10^{-6} lifetime risk to the U.S. population implies 236 additional deaths over a 70-year interval. While this may seem to be an acceptable level—that is, a trivial level of risk in comparison to the 2 million annual U.S. deaths—to require that this same level of risk also be applied to populations irrespective of their size would be fundamentally wrong. It is wrong from two points of view. First, it seems to suggest that if 236 deaths amongst the entire population is inherently acceptable, then if the population at risk is considerably reduced it remains an acceptable risk. Thus, if a limit of 236 deaths is selected as the criterion of acceptable risk, we would have to affirm that were 236 deaths to result from some chemical intermediary never released as a trace contaminant into the environment and resulting in cancer deaths totally confined to a single factory, this would also be acceptable. For if all the deaths were limited to the factory work force, there would be no other deaths in the U.S. population and it would still be 236 deaths per 236 million people, a 10^{-6} risk. This is the rate we have considered to be acceptable. But in fact, this is absurd: It is unacceptable to me, to society, and certainly to the factory workers who are at risk. Is this extreme example merely a straw man or does it point to an inherent weakness in setting a rate of risk that is acceptable without at the same time specifying additional constraints? I suggest that it demonstrates that simply using a specific number of deaths as the criterion of acceptable risk, without relating it to the size of the population, is an inadequate and incomplete approach. An important ancillary consideration, related to the size of the exposed population and similar in force to the voluntary/involuntary and old/new dualities, is the random, dispersed, and accidental versus nonrandom, localized, and intentional duality of the risk. The size of the population at risk is an objective quantity; the randomness that seems to be related, a subjective one. Two hundred and thirty-six random U.S. deaths are acceptable, but the same number, if localized and nonrandomized, is considered to be a disaster.

The second reason it is wrong not to consider population size when setting acceptable levels of risk becomes evident when the 10^{-6} individual rate of risk is considered. A rate of this magnitude (or arguably, of magnitude 10^{-5}) is a rate that is realistic and prudent (as actual risk numbers confirm) when the population at risk is very large, 236 million in the example with which we are concerned. But if it applied as a rate of risk to a discrete factory or community population that is uniquely at risk from a carcinogenic chemical, its consequences become extreme: A myriad of society's essential activities would have to cease. For example, the sawdust created when sawing several 2 by 4s each week probably places the carpenter at a lifetime risk of at least 10^{-6} for cancer of the oral cavity and pharynx. Certainly the X-ray technician, the white-collar worker in a home or office constructed with the ubiquitous binding agent urea-formaldehyde, and the short-order cook exposed to benzopyrene in the smoke from charcoal-broiled hamburgers are each at an individual cancer risk considerably higher than 10^{-6}. Indeed, even the farmer in an agricultural society is at a 10^{-3} to 10^{-4} risk of malignant melanoma from pursuing his trade in the sunlight. The 10^{-6} criterion may be appropriate when many millions are at risk, but to enforce such a level of risk when exposed populations are small is just not a realistic option.

These examples also emphasize an important distinction between de minimis risk and acceptable risk: None of these examples of risk is really trivial (although a $10^{-3} = 0.001$ incremental lifetime risk for a single individual may certainly be so viewed!), and we acknowledge that such risks are inexorably associated with living and inherently unavoidable. Indeed, refraining from these activities generally results in the substitution of other commensurate risks.

The suggestion that the size of the population or the population density is irrelevant when setting acceptable levels of risk can be faulted from a third perspective. If it were so that individual risk in areas of low population density and high population density need not differ, there would be no objection based upon risk analysis to siting nuclear power plants within highly populated metropolitan centers. In fact, neither those opposed to nuclear power nor representatives from the nuclear power industry itself would seriously consider this option.

Two conclusions may be drawn from the foregoing: First, the size of the population at risk is a pertinent consideration in determining a rate of acceptable or de minimis risk. Second, neither a risk relationship that varies at an inverse population rate [$R_L = M/P$ where M is total lifetime mortality (236 in the above example), P is the population at risk, and R_L is lifetime risk, see Figure 1] nor one that is independent of population ($R_L = 10^{-6}$, Figure 1) achieves a result that is reasonable from both subjective and objective perspectives. To be more explicit: For the entire U.S. population ($P = 236$ million), the two expressions, by an appropriate choice of M, can give results that are consistent and reasonable. But then for small populations (e.g., 100) the expressions lead to different results, each of which is untenable.

A compromise between these two risk formulations can avoid this weakness. The two expressions for lifetime risk R_L given above are shown in Equations (1) and (2). Multiplying the two together gives Equation (3). The square root of Equation (3) yields Equation (4), the risk-population equation. The procedure is equivalent to taking the geometric mean of Equations (1) and (2), and results in a relationship that avoids the practical and conceptual weakness of both these formulations. The risk-population equation is plotted in Figure 1.

$$R'_L = 236/P \quad (M' = R'_L P = 236) \tag{1}$$

$$R''_L = 10^{-6} \quad (M'' = R''_L P = 10^{-6}P) \tag{2}$$

$$R^2_L = 236 \times 10^{-6}/P \quad \text{where} \quad R^2_L = (R'_L)(R''_L) \tag{3}$$

$$R_L = 0.015/P^{1/2} \quad \text{(the risk-population eq.)} \tag{4}$$

If risk-population data points fall below the Equation (4) line, they are to be considered to represent de minimis or acceptable carcinogenic risks. Data points falling above the line are not to be so considered.

Using Equation (4) with $P = 100$, the lifetime risk is 1.5×10^{-3} and the annual risk is 2.14×10^{-5}. This value is virtually the same as the standard deviation of the quadratic equation that can best be fitted to the 1982–1983 monthly mortality rate data

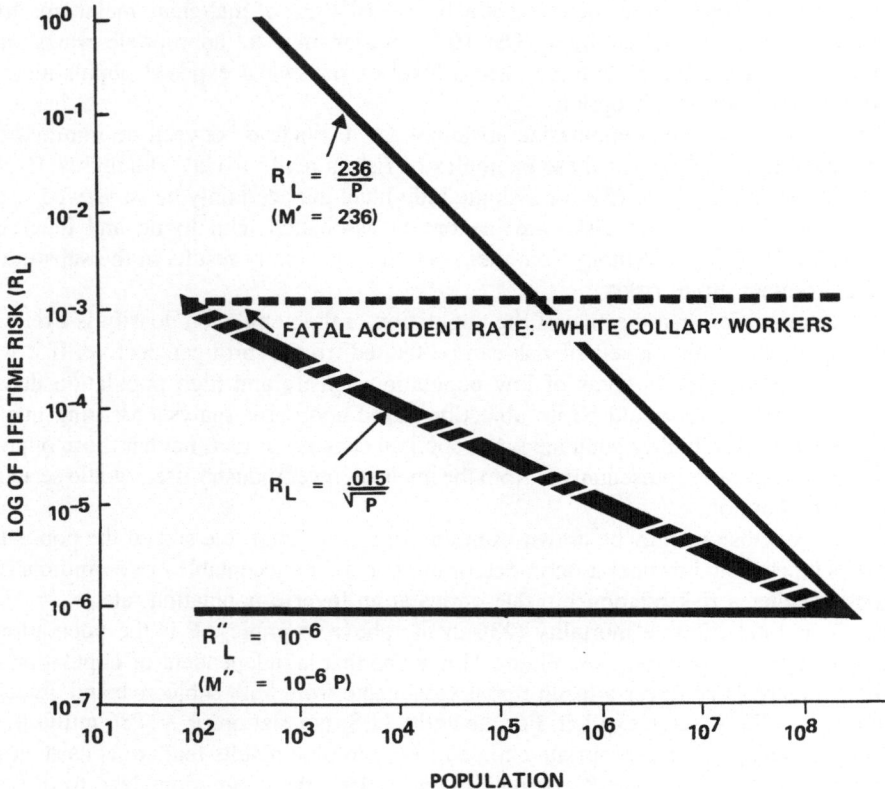

Figure 1. Two rates of lifetime risk (R'_L and R''_L) as a function of the size of the population at risk are presented. R''_L shows a rate of (individual) risk that is equal to 10^{-6} and is independent of population. R'_L shows a total (population) risk for which the total lifetime mortality is equal to 236 and that is independent of the size of the population. The text suggests that neither constitutes an acceptable approach to the determination of a de minimis risk for carcinogenic chemicals but that their geometric mean, given by R_L, provides an approach that is attractive in terms of criteria of both objective and subjective measures.

(2.11×10^{-5}) shown in Table 1. It is also nearly equal to the lowest annual fatal occupational rate from accidents. This rate (2.35×10^{-5}), equal to a lifetime rate of 1.6×10^{-3}, occurs in the "finance, insurance and real estate" occupational category.[11] It is selected as an appropriate reference point for a de minimis risk when $P = 100$ because it is considered improbable that a prospective employee would be reluctant to accept a job in this employment category because he or she considered the risk from fatal accidents to be unacceptably high. Were one seeking employment with safety from fatal accidents as the sole criterion of job acceptability, one would *seek out* just such job categories. Finally, the risk-population equation is deemed appropriate only for populations that number 100 or more because empirical data suggest that smaller populations are not really relevant in the real world in which environmental and occupational carcinogens almost invariably expose groups of more than 100 people. Thus, a recent NIOSH analysis[12] of 75,000 potentially toxic chemicals that are used in commerce, industry, and agriculture has shown that in about 90% of the instances in which a worker is exposed to a single toxic chemical, 7500 other U.S. workers are also exposed to the same chemical. More than 99.9% of workers who are exposed to a toxic chemical—not all of which are carcinogenic—are in a cohort that has at least 100 members that also experience this same exposure. Put somewhat differently, fewer than 2 of every 10,000 workers who are exposed to a toxic chemical are exposed under conditions in which fewer than 100 other workers are also exposed to the same chemical (although not necessarily in the same factory or location).

It is of interest to examine recent historic risk and population data from the U.S. Environmental Protection Agency and to compare the regulatory decisions based upon these data with the results of the foregoing analysis. Figure 2 and Table 2 present these data. Solid squares represent carcinogenic hazards for which regulations have been pro-

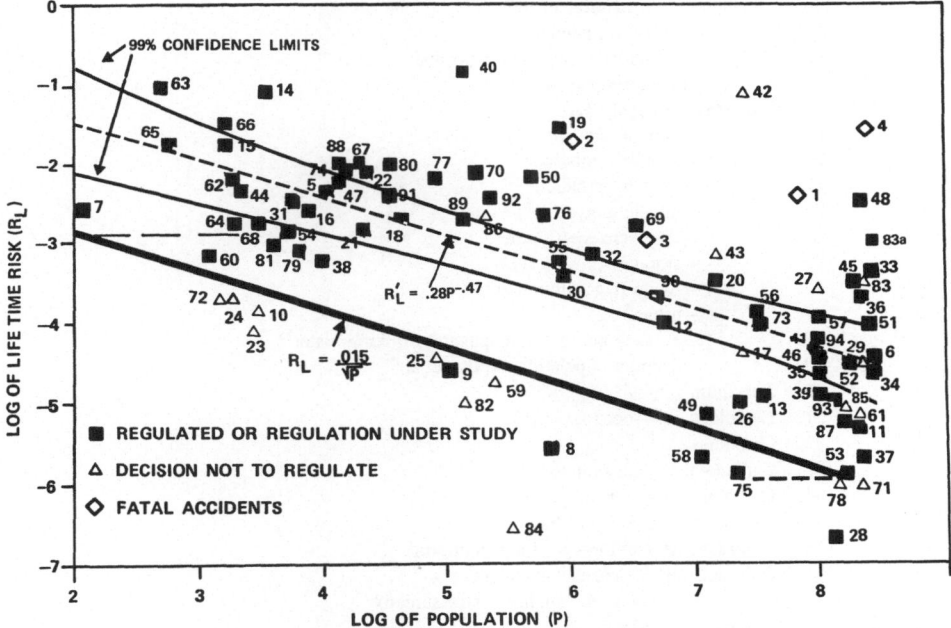

Figure 2. Lifetime cancer mortality risk R_L is shown as a function of the magnitude of the exposed population. The curve R_L is derived in the text. Risks that are less than R_L are considered by the author as constituting de minimis risk. In addition to this line, 94 data points (solid squares, open triangles, and open diamonds) indicate lifetime estimates of chemical and accident fatalities. The placement of these points is determined by the risk and the size of the exposed populations, as estimated primarily by EPA. The squares represent chemicals actively under study for regulation (or for which regulations have already been promulgated). The triangles represent the decision not to regulate the chemicals so designated. The diamonds provide fatal accident rates. The dotted line represents the best possible straight line that can be drawn through the squares. This suggests that, in general, EPA manages carcinogenic risk as if the size of the exposed population were relevant (either implicitly or explicitly, either as a dependent—or surrogate—variable or as an independent variable) to the risk-setting and risk decision-making process. (The indoor radon datum point (83a) is shown in the figure but is not included in the calculation to obtain R_L. Nonregulated data points and fatal accident points are also excluded.)

Table 2. Key to Figure 2[a]

1.	Accident, fatal—private sector 1982
2.	—mining
3.	—finance, insurance, and real estate, 47 years (18–65)
4.	—all, 1982 rate for 70 years
5.	Acrylonitrile[8]
6.	Alachlor—dietary[17]
7.	—flaggers[17]
8.	—farmers[17]
9.	—ground applicators[17]
10.	Amitraz—apple and pear sprayers[4]
11.	—apple and pear consumers[4]
12.	—apple, dietary[9]
13.	—pears, dietary[9]
14.	Arsenic—copper smelters: high[9]
15.	—copper smelters: low[9]
16.	—glass manufacturing[9]
17.	—Inorganic, neighboring population average exposure[4,9]
18.	—maximum exposure[4]
19.	Asbestos—occupational[9]
20.	—school; students and teachers[10]
21.	Benzene—fugitive emission[9]
22.	—coke by-product[7,9]
23.	—maleic anhydride[9]
24.	—ethylbenzene/styrene[9]
25.	—storage[9]
26.	—Stage II gasoline market[7]
27.	—urban[6]
28.	—average population exposure: drinking water[26]
29.	—average population exposure: air[4]
30.	Beryllium[3]
31.	Butadiene,1,3—occupational[20]
32.	Cadmium[11]
33.	Captan—food consumption[18]
34.	Captofol—food consumption[19]
35.	Carbon tetrachloride—urban[6]
36.	Chlordane/heptachlor—food consumption[4]
37.	Chlorobenzilate—citrus consumers[4]
38.	—citrus applicators (assumed)[4]
39.	Chloroform—urban[6]
40.	Chromium—vicinity of large point sources[11,15]
41.	—urban[11]
42.	Cigarette smokers—male
43.	Coke ovens—averge exposure for U.S. population at risk[4]
44.	—occupational[4]
45.	Daminozide—food consumption[22]
46.	1,2-dichloroethane—urban[6]
47.	Ethylene dibromide—occupational[9]
48.	—immediate postregulatory dietary[9]
49.	Ethylene dichloride[1]
50.	Ethylene oxide[11]
51.	Folpet—food consumption[23]
52.	Formaldehyde—urban ambient[9]
53.	—production use release[9]

Table 2. (Continued)

54.	—resin manufacturing workers[9]
55.	—apparel workers[9]
56.	—mobile homes[10]
57.	—non-urea/formaldehyde homes[10]
58.	Lindane—shelf paper[3]
59.	—livestock applicators[3]
60.	—pecan applicators[3]
61.	—food[3]
62.	—indirect occupational exposure[9]
63.	—direct occupational exposure[9]
64.	—non-production workers (occupational)[9]
65.	—manufacturing workers[9]
66.	—processing workers[9]
67.	—all workers: OTS-based[9]
68.	—all workers: TLV-based[9]
69.	Nickel[11]
70.	Nitrosamines—occupational exposure from metal-working fluids[21]
71.	NTA—public drinking water[4]
72.	—formulators (occupational)[4]
73.	PCB—dietary fish[10]
74.	Pentachlorophenol—applicators/workers[25]
75.	—air[2]
76.	Radiation, ionizing—all workers in medicine and industry[12]
77.	—power reactor workers[12]
78.	—coal-fired boilers[16]
79.	Radionuclides—DOE facilities[16]
80.	—uranium mines[16]
81.	—elemental phosphorus plants[16]
82.	—phosphate industry[16]
83.	Radon—drinking water[5]
83a.	—indoor[27]
84.	Styrene monomer—occupational[2]
85.	Tetrachloroethylene—urban population[14]
86.	—dry cleaners[14]
87.	Trichloroethylene—urban air[14]
88.	Uranium mill tailings—inactive sites[9]
89.	—active sites[9]
90.	Vinyl chloride—average exposure for population: air[4]
91.	—maximum exposure (occupational)[4]
92.	—workers (occupational)[11]
93.	—average exposure to population: water[26]
94.	Volatile synthetic organic compounds[24]

[a] The values provided in the references to this table reflect current state-of-the-art estimates of risk and population at risk. They are subject to revision and improvement, and their inherent limitations should be borne in mind.

mulgated or are under active consideration. (If the hazard has been regulated it is the preregulated risk that is shown.) Open triangles represent hazards for which no regulation was deemed to be necessary. The linear equation (using log–log coordinates) that best fits the data is shown by the thin dotted line. Its slope ($= -0.47$) is very nearly the same as the slope ($= -0.50$) of the population-risk equation (Eq. 4) also shown in the figure. Although the lines are nearly parallel, the line generated from the data is displaced almost one and a half orders of magnitude above the risk-population equation. This implies that these chemicals lie above the de minimis or acceptable risk level and should (from the perspective of this analysis) be regulated. That they are under consideration for regulation (or already regulated) is therefore consistent with our analysis. The fact that the slopes are so nearly the same seems also to suggest that it is recognized, although perhaps only implicitly, by the EPA's risk managers, that the size of the population at risk is a valid factor that has to be considered in the regulation of chemical carcinogens. Also consistent with the analysis is the fact that 10 of the 15 chemicals or data points that fall below the line are not being considered for regulation. The analysis suggests that 9 of the 76 chemical data points that lie above the line, although not now being considered for regulation, in fact do present a sufficiently high risk to a sufficiently large population to warrant regulation.

Considerable progress has been made in recent years in the analysis of risks that arise from human exposures to carcinogenic hazards. The management of these risks has been based, first and foremost, upon the empirical exposure data and the carcinogenic potency of the chemicals of concern, each considered within the general context of today's scientific and medical understanding of the carcinogenic process.

But in addition to considerations such as those related to economic costs and benefits, which we have neglected in this discussion, risk management has also always incorporated into the regulatory decision-making process (generally in an *ad hoc* way) the real-world pressures, influences, and concerns expressed by the public at large that derive from their perceptions of these risks. The general population's assessment of these risks often involves subjective factors that are to a considerable degree culturally determined. These, in turn, often are further modulated by the self-interest of particular social subpopulations such as workers, environmentalists, or industrial managers.

I have argued above that the decision-making process undertaken by risk managers *inherently* incorporates aspects that are both objective and subjective. To determine the "real" level of acceptable or de minimis risk and then to adjust (sometimes the self-serving term "fine tuning" is used to describe this process) this level so that it is compatible with people's subjective perceptions and with their commonsense notions of what a reasonable level really is, is a stepwise or sequential approach that is *ad hoc* and less than satisfying from a conceptual perspective. The subjective/objective duality of risk management, if considered in parallel from the very inception of the risk management process, is more likely to result in methodological approaches to the management of chemical carcinogens that are more consistent and reasonable than otherwise would be the case.

The sequential approach initially disregards all data that are not objective. Hazard identification, carcinogenic potential, and quantifiable human exposures are the data to be considered, modulated only by the objective dictates embodied in legal statute. But we recognize that in the real world where human tears, anguish, and fear are inseparable from statistical printouts, the risk manager ultimately must adjust such objective analyses by considering the human dimensions of the problem. This recourse to common sense, be it perceived as pragmatism, political savvy, or an instance of democracy at work, is the second essential step of the risk manager's job and is probably the main feature of this discipline that distinguishes it from the risk analyst's task. But because this accommodation with the constraints of reality as sensed and expressed by the body politic is usually done in an *ad hoc* and *ex post facto* fashion, it often seems to conflict with the results of the objective analysis of risk. This is not to suggest that this finetuning that results from a multiplicity of social pressures invalidates the analysis of an acceptable risk level for the particular hazard under review. To the contrary, in the final analysis it legitimizes it. What it does suggest, however, is that this pragmatic approach fails to extract from each particular instance of risk management more generalized insights that can then serve as guidelines to inform the practice of future regulatory decision making.

The alternative to this sequential two-step approach, one that avoids these weaknesses, has been proposed in this chapter. It is an approach that would integrate, at the very earliest stages of the risk manager's efforts, the body politic's subjective notions of acceptable risk and the strictly analytical and objective description of risk. With this objective in mind, the influence of the size of the population at risk (and the related random/nonrandom quality of causation) on our notions of acceptable risk levels has been considered. Finally, a level of acceptable or de minimis risk has been suggested that strikes a balance between the results of a largely objective analysis of acceptable risk and what I see as being notions of acceptable risk, more subjective in character, that emerge from deeply embedded cultural attitudes.

It is considered significant that this integrated one-step approach results in levels of acceptable or de minimis risk that are compatible with recent decisions by the EPA to study, with a view toward regulating, the carcinogenic risks associated with several specific chemicals.

REFERENCES

1. Covello, V. T., Flamm, W. G., Rodricks, J. V., and Tardiff, R. G., eds., *The Analysis of Actual Versus Perceived Risks*, Plenum Press, New York, (1983).
2. Fischhoff, B., Lichtenstein, S., Slovic, P., Derby, S., and Keeney, R., *Acceptable Risk*, Cambridge University Press, New York, (1981).
3. Fischhoff, B., "Judgmental Aspects of Risk Assessment," in *Risk Assessment and Risk Assessment Methods: The State-of-the-Art*, National Science Foundation (NSF/PRA-84016), Washington, D.C., (December 1984).
4. Litai, D., Lanning D. D., and Rasmussen, N. C., "The Public Perception of Risk," in Covello, V. T., et al., eds., *The Analysis of Actual Versus Perceived Risks*, 235– 249, Plenum Press, New York, (1983).
5. Adler, H. I., and Weinberg, A. M., "An Approach to Setting Radiation Standards," *Health Physics 34*, 719–720, (1978).
6. Whipple, G. H.,"A Practical Threshold for Radiation," June 10, 1980, cited by Joyce P. Davis in *The Feasibility of Establishing a "De Minimis" Level of Radiation Dose and a Regulatory Cut-off Policy for Nuclear Regulation*, General Physics Corp., Columbia, Md., (December 31, 1981).
7. U. S. Dept. Health, *Vital Statistics of the U. S. II, Part A.*
8. National Center for Health Statistics, *Monthly Vital Statistics Report 32* (13), p. 4.
9. Comar, C. I., "Risk: A Pragmatic De Minimis Approach," *Science 203*, 319, (January 26, 1979), reprinted in this volume.
10. Doniger, D. D., Statement before the Subcommittee on Oversight and Investigations Committee on Energy and Commerce, Washington, D.C., (November 7, 1983).
11. U.S. Department of Labor, Bureau of Labor Statistics, *Occupational Injuries and Illnesses in the United States by Industry, 1982.* Bulletin No. 2196, Table 8, 1982 data.
12. National Institute for Occupational Safety and Health, *National Occupational Exposure Survey, 1981– 1983*, Cincinnati, O., (1984).
13. Milvy, P., "A General Guideline for Management of Risk from Carcinogens," *Risk Analysis 6*, No. 1, 69–80, (1986).

REFERENCES FOR TABLE 2

1. Suta, B., *Assessment of Human Exposures to Atmospheric Ethylene Dichloride* SRI International, (May 1979).
2. EPA, Office of Policy Analysis, *Unit Risk Estimates For Toxic Air Pollution*, Office of Air Quality Planning and Standards, *Maximum Exposure Levels and Population Totals*, (1984).
3. EPA, Lindane PD-4 (draft), (August 1983).
4. Anderson, E. L., "Quantitative Approaches in Use to Assess Cancer Risk," *Risk Analysis 3*, No. 4, 277–295, (1983).
5. Cothern, R. C., et al., *Development of Quantitative Estimates of Uncertainty in Environmental Risk Assessment When the Scientific Data Base is Inadequate* (Draft), Office of Drinking Water, EPA, Washington, D.C.
6. Bussard, D., Memorandum dated 3/15/84, EPA, Washington, D.C.
7. Gorman, T., NESHAP briefing paper, OPPE, EPA, (1984).
8. Office of Air Quality and Standards, *Need for Regulation of Coke Oven Emissions and Acrylonitrile Under CAA*, briefing paper, EPA, Washington, D.C., (March 1984).
9. Dobkowski, D., Memorandum to A. Jennings dated 4/3/84, Acting Director Statistical Policy Division, EPA, Washington, D.C.
10. Chemical Coordination Staff for the Six Month Air Toxics Study "Acceptable Risk Levels and Federal Regulations . . .," EPA, Washington, D.C., (May 1984).
11. Haemisegger, E., Jones, A., et al, *The Air Toxics Problem in the United States: An Analysis of Cancer Risks for Selected Pollutants*, EPA (Final Agency Internal Review); Washington, D.C., (May 1985).
12. Kumazawa, S. et al., *Occupational Exposure to Ionizing Radiation in the United States*, EPA, Washington, D.C., (March 1983).
13. Britton, B., *Risk Characteristics for Various Pollutants Regulated or Being Considered by EPA Program Offices*, Chemical and Statistical Policy Division, EPA, Wash., D.C., 1985.
14. Milvy, P., *Estimates of Cancers from Perchloroethylene (PCE) Exposure (4/3/84) and Health Assessment Document for Tetrachloraethylene*, EPA, Washington, D.C. (December 1983).
15. EPA-OHEA, Health Assessment Document for Chromium, 7/83 draft.
16. Office of Radiation Programs, *Background Information; Final Rules for Radionuclides*, 11, (October 23, 1982).
17. EPA, Draft Alachlor PD-1 (12/4/84), 54.
18. EPA, Draft Captan PD-2/3 (2/5/85), 11–68.
19. EPA, Captofol PD-1, (December 1984).
20. EPA, Office of Toxic Substances, *Assessment of Cancer Risk to Workers, Exposure to 1,3-Butadiene in Plants Producing Synthetic Rubber, Plastics and Resins*, (November 21, 1984).
21. Preliminary Economic Analysis of Proposed Regulations for the Use of Nitrites in Metalworking Fluids; PHD, Inc., (October 1984).
22. EPA registration standard, 21, (June 84).
23. EPA, Preliminary Folpet Risk Assessment Briefing Paper for S.I.S., OPP, (1985).
24. EPA, Regulatory Impact Analysis, Regulatory Flexibility Analysis and Paperwork Reduction Act Analysis for Proposed Regulations to Control Volatile Synthetic Organic Chemicals (VOCs) in Drinking Water (EPA -570/9-85-004), (Calculated from pages I-5 and IV-8) (May 1985).
25. EPA, Office of Pesticide Programs Position Document 2/3 on Wood Preservatives, 364, 582, 589, (1984).
26. Cothern, R. C., Coniglio, W. A., and Marcus W. L., *Techniques for the Assessment of the Carcinogenic Risk to the U.S. Population Due to Exposure from Selected Volatile Organic Compounds from Drinking Water Via the Ingestion, Inhalation, and Dermal Routes* (EPA-570/9-85-001), (July 25, 1984). (Calculations based on the multistage model for extrapolation of risk to low dose).
27. Nero, A. V. Jr., "Risk and Policy Implications of Indoor Exposure to ^{222}Rn Decay Products and Other Air Pollutants." Paper presented at the 1985 Annual Meeting of the Society for Risk Analysis, Alexandria, Va.

8

Carcinogenic Potencies and Establishment of a Threshold of Regulation for Food Contact Substances

W. Gary Flamm, L. Robert Lake, Ronald J. Lorentzen, Alan M. Rulis, Patricia S. Schwartz, and Terry C. Troxell

INTRODUCTION

This chapter presents a compilation of data on carcinogenic potencies in the form of a probability distribution and describes how, in this form, the data could be used as a scientific foundation for making regulatory judgments about the potential risk from substances that migrate to food from packaging and other materials that come into contact with food. The idea of establishing a threshold of regulation based on the concept of a de minimis level of migration of potentially carcinogenic substances into food is discussed as a possible regulatory approach that could use the type of data presented. The discussion does not represent Food and Drug Administration (FDA) policy, however, and it does not address the possible use by FDA of a de minimis risk approach to determining the status of certain color additives currently being evaluated by the agency.

BACKGROUND

Substances that come into contact with food through packaging, handling, or processing are regulated by FDA under the food additives provisions of section 409 of the Federal Food, Drug, and Cosmetic Act (the act). The basis for regulating food contact

W. Gary Flamm, L. Robert Lake, Ronald J. Lorentzen, Alan M. Rulis, Patricia S. Schwartz, and Terry C. Troxell • Center for Food Safety and Applied Nutrition, Food and Drug Administration, Department of Health and Human Services, Washington, D.C. 20204.

substances as food additives is found in the food additive definition in section 201 (s) of the act. That definition states that a food additive is an intentionally used substance that "results or may reasonably be expected to result, directly or indirectly, in its becoming a component or otherwise affecting the characteristics of any food. . . ." Because substances that come in contact with food ordinarily migrate into food to some extent, they are regulated by FDA in response to petitions from the industry to approve their uses as indirect food additives.

FDA took the position that any level of migration of any component of a food-contact material required the material to be regulated as a food additive before the ruling in case of *Monsanto* v. *Kennedy*.[1] The *Monsanto* case dealt with the regulation of acrylonitrile copolymer bottles intended for beverage use. The U.S. Court of Appeals (D.C. Circuit) decided that the FDA had more discretion than it apparently believed, and that the agency could determine that a substance in contact with food presented a de minimis situation.

The court in *Monsanto* did not define what a de minimis situation was, but it did indicate, among other things, that FDA had considerable discretion in determining when a food-contact material should be regulated as a food additive, even discretion to assure a wide margin of safety. The notion that some food-contact materials could be permitted without a food additive regulation gives rise to the possibility of establishing a threshold of regulation for those migrants that clearly would not pose a public health concern.

THRESHOLD OF REGULATION

FDA has for some time been considering what a threshold of regulation situation might be. The idea is to determine the possible circumstances for permitting the use of food-contact materials without the necessity of issuing a food additive regulation.

Although this discussion is not intended to deal with policy, the foregoing discussion was necessary to an understanding of the context in which the data that are to be presented are being considered. The remainder of the chapter will focus on certain scientific problems in determining the level of human dietary exposure resulting from migration that might be used in setting a threshold of regulation for food-contact substances.

It would, of course, be convenient if FDA could define a threshold or regulation in terms of a single achievable level of migration. This would provide FDA and the regulated industry with a simple means of determining whether a substance used in a particular way in contact with specific foods requires a food additive regulation—or whether the uses of a substance could be permitted by some other means. There are, however, difficulties with developing such a simple approach that need to be addressed.

One problem is that the toxicity of chemicals varies widely. In fact, carcinogenic potencies, according to Gold et al.,[2] are known to vary on a weight basis, by a range of 1 billionfold. This means that even a low level of exposure to a substance at the high end of the cancer potency range would be a matter of public health concern because an extraordinarily potent carcinogen would be expected to present a significant risk unless the total exposure were extraordinarily low. For example, while one part per million of a particular substance in food may represent a very small risk for a weak carcinogen, this same level of exposure could create a public health problem for a very potent carcinogen. Consequently, if risk assessment is to be used to determine a threshold of regulation, and many believe that risk should be the deciding factor for carcinogens, or potential carcinogens, potency must be a critical consideration. Further, if actual potency is unknown, then the range of possible potencies becomes important.

Another problem with a single-threshold level of migration is that use patterns of food packaging materials vary considerably. This means that the same level of migration from different packaging materials corresponds to different levels of human exposure, depending on how widely the packaging material is used. Consequently, FDA's safety evaluations of food contact materials have historically relied heavily on an assessment of probable exposure, using migration data and market survey information.

CANCER POTENCIES

The major toxicological concern for food packaging migrants has been cancer induction, because the prevailing view is that most carcinogens do not have "no effect" levels. The critical technical problem in setting a threshold of regulation level is how to deal with the wide range of cancer potencies, assuming that long-term feeding studies to determine the actual cancer-causing potential for hundreds of food-contact substances will not be conducted.

One approach is to examine the data from long-term feeding studies that currently exist on a wide range of carcinogens, and then to see what conclusions might be derived from these data. The idea is to determine how low a level of exposure is required to ensure that the upper bound of lifetime risk from most any ingested carcinogen is well below some specific level of risk deemed to be acceptable.

In Figure 1 a probability distribution of the potencies of known carcinogens is presented. In constructing this distribution we utilized a recent compilation of hundreds of carcinogenic potencies by Gold et al.[2]

The solid line seen in Figure 1 represents the (nonlinear least squares) best fit to the observed distribution calculated using a Gaussian functional form. A visual comparison of the observed histogram distribution to the calculated curve demonstrates that the observed distribution of carcinogen potencies is surprisingly Gaussian. Hence, for simplicity we shall, throughout the remainder of this chapter, allow the distribution of carcinogen potencies, and any exposure distributions derived from them, to be represented by the appropriate Gaussian curves instead of histograms.

In order to make use of this potency distribution in addressing questions of low-level exposures, it is useful to transform the *potency* distribution into a corresponding *exposure* distribution. This may be done by establishing a given risk standard, say 10^{-8} or 10^{-6} per lifetime, and using the simple assumption that low-dose risk equals potency times exposure (where, for the purposes of this discussion, potency is taken to be the slope of a straight line connecting zero risk/zero exposure to a point corresponding to a risk of 0.5 at the TD50 dose of Gold et al.[2] With such a "risk-equivalent" exposure distribution, it is possible to analyze the consequences of relating a given exposure level to a possible threshold of regulation for migrants. For example, the choice of any dietary

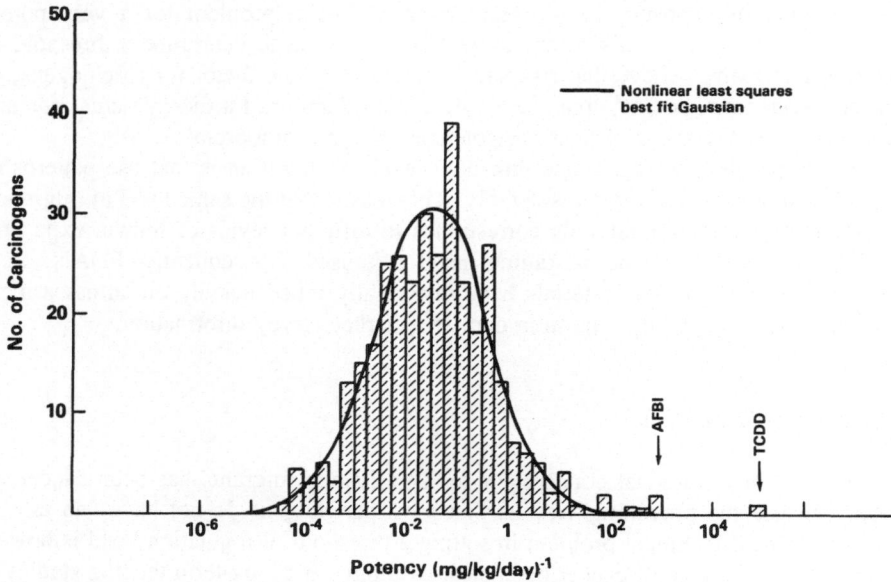

Figure 1. Histogram and nonlinear least squares best fit Gaussian of the semilogarithmic potency distribution of 343 carcinogens. Using data from Gold et al., 343 compounds/bioassays were chosen from a total of 770 compounds and 3000 carcinogen bioassays. The selection criteria were (1) oral route of administration, (2) p value < 0.01 for increased incidence of animals with specific neoplasms, (3) most sensitive species/sex/organ site combination. *AFB1* and *TCDD* refer to Aflatoxin B1, and 2,3,7,8 Tetrachlorodibenzo-p-dioxin, respectively. The arrows indicate the approximate position for these compounds in the distribution.

exposure level such as one part per trillion (ppt) will, when placed in the context of the "risk-equivalent" exposure distribution, enable a determination of how many or what percentage of total carcinogens could exceed the selected risk standard.

For instance, as shown by the arrow in Figure 2, 1 ppt corresponds to a dose of between 10^{-7} and 10^{-8} mg/kg per day. Roughly a quarter of the carcinogens included in the analysis exceeded a 10^{-8} risk at the 1-ppt level in the entire diet.

Figure 3 illustrates the same principle using a plot of the dose required to produce a 10^{-6} risk. The same 1-ppt cutoff would exclude almost all risks of 10^{-6} or greater from substances migrating to food at that level in the diet. Clearly, any risk level could be chosen and a similar analysis undertaken to determine how and with what confidence the chosen carcinogenic risk could be excluded given the use of a simple and convenient parameter such as concentration in the total diet.

It is obvious that the validity of this approach depends at least upon two critical assumptions: (1) that the carcinogens whose potencies have been considered here effectively represent the universe of chemical carcinogens present in food-contact materials and that the discovery of new carcinogens will not change significantly the distribution presented here, and (2) that the linear relationship between dose and risk used for transforming carcinogen potencies to doses corresponding to upper-bound risks represents a reasonable approximation.

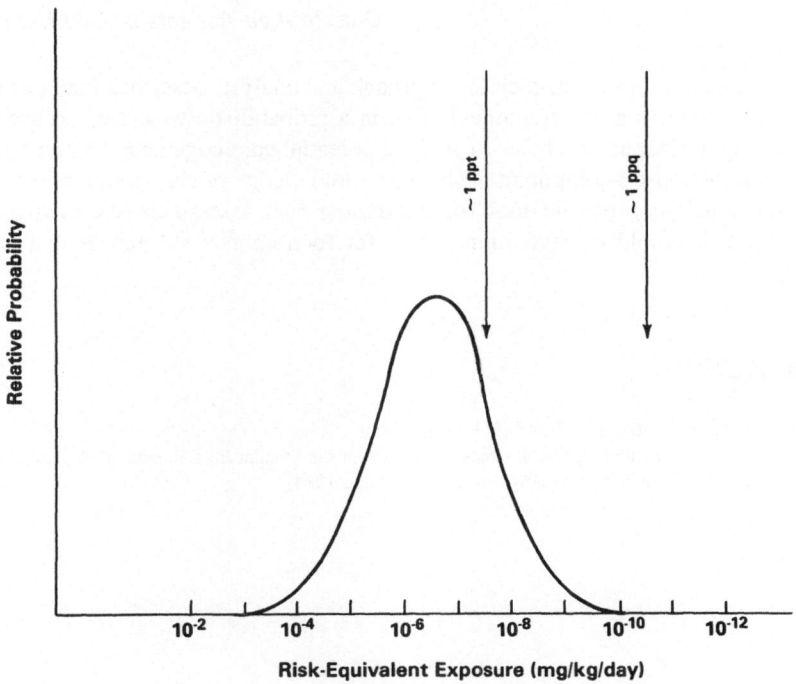

Figure 2. Semilogarithmic exposure distribution calculated from the potency distribution in Figure 1 with a lifetime risk standard of 10^{-8}.

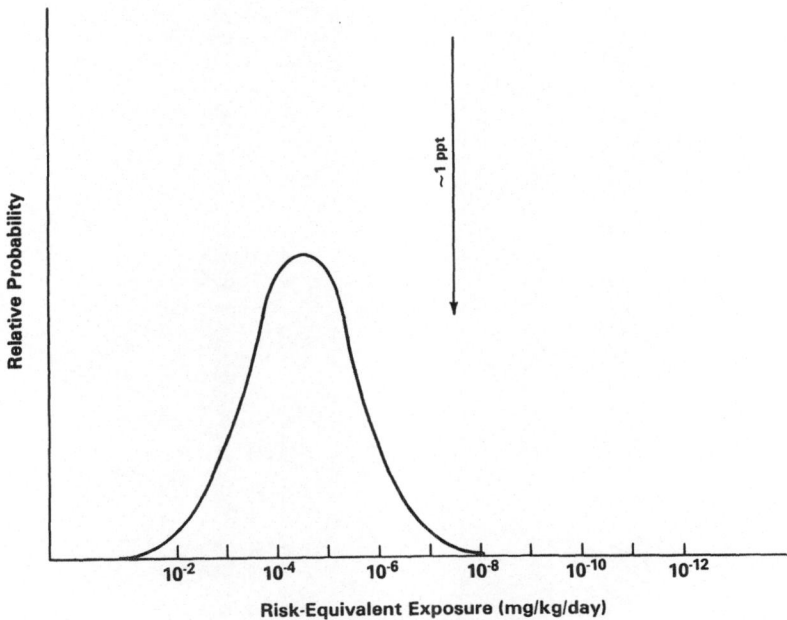

Figure 3. Semilogarithmic exposure distribution calculated from the potency distribution in Figure 1 with a lifetime risk standard of 10^{-6}.

If both assumptions hold, then the approach and analysis described here can be used to consider the question of carcinogenic risk in a probabilistic way, i.e., to use present knowledge of carcinogen potencies to analyze potential carcinogenic risks globally rather than on a compound-by-compound basis. Thus, knowledge of carcinogen potencies provides FDA a useful scientific tool for addressing risk assessment issues involved in establishing a threshold of regulation policy for food-contact substances that have not been tested.

REFERENCES

1. *Monsanto* v. *Kennedy*, 613 F. 2d 947 (D.C. Cir, 1979).
2. L. S. Gold, et al., "A Carcinogenic Potency Data Base of the Standardized Results of Animal Bioassays," *Environmental Health Perspectives 58*: 9–314 (December 1984).

III

Regulatory Applications of De Minimis Risk

9

Implications of De Minimis Risk Concepts for OSHA

Raymond E. Donnelly

The legal framework provided by the Occupational Safety and Health Act of 1970 and by subsequent Supreme Court decisions interpreting that law are a logical starting point for discussion of de minimis risk concepts and their application in OSHA. If we work initially in that *legal* dimension, the social, ethical, and practical dimensions will quickly become apparent. In these remarks, I hope to point out where de minimis concepts have been successfully applied already, and to point toward program areas where they may apply in the future.

First, the OSHAct. Section 6(b) of the act states, in part,

> . . . in promulgating standards dealing with toxic materials or harmful physical agents . . . [the agency] shall set the standard which most adequately assures, to the extent feasible, on the basis of the best available evidence that no employee will suffer material impairment of health or functional capacity even if such employee has regular exposure to the hazard dealt with by such standard for the period of his working life.

The plain sense of this language is that OSHA is to be conservative in its approach to standard setting, and that any *material impairment of health or functional capacity* is to be prevented. Given the wide range of workplace hazards and their potential harm to health, this provision might be interpreted to give the agency very little room to describe any risk as *de minimis* if it could lead to material impairment of health or functional capacity.

The views expressed here do not necessarily reflect the policies or position of OSHA.

Raymond E. Donnelly • Occupational Safety and Health Administration, U.S. Department of Labor, Washington, D.C. 20210.

But that provision is not absolute. Section 3 of the act calls for measures that are "reasonably necessary and appropriate," indicating the need for administrative judgment in the standards process.

In the benzene case [100 s.ct.2844 (1980)], the Court ruled that in order for an OSHA standard to be valid, the risk in question must be a significant one, and it must be likely to be reduced or eliminated by the regulatory measures OSHA proposes. The Court indicated that "significant risk determination is not a mathematical straitjacket and that OSHA is not required to support its finding that a significant risk exists with anything approaching scientific certainty." This ruling, on the one hand, constrained the agency to support its risk determinations with reasonable evidence but, on the other hand, left it free to leave unregulated other risks for which the evidence is less complete or less persuasive and, especially, those risks which are considered to be of low probability or low consequence. Thus, from the benzene case, two principles were established for OSHA standard setting: (1) There must be a demonstration of significant risk. (2) It must be shown that the measures proposed in the standard will reduce the risk.

In addition to these, a third principle was derived from the cotton dust case, in which the Supreme Court upheld OSHA's standard regulating occupational exposure to cotton dust (101 S.Ct. 2478 [1981]). In that case, it was established that the measures proposed must be feasible economically and technologically for the affected industry as a whole. This has, in practice, led to the articulation of a fourth principle, that OSHA regulations must be cost-effective, thus providing protection in the most economically efficient way.

Having sketched in bare outline some of the legal framework for OSHA standard setting, I feel that it is helpful to broaden the perspective to include OSHA enforcement strategies—that is, what OSHA actually does with the standards that it has on its books. It is in this area that some de minimis concepts are already in place and functioning.

In many ways, the de minimis concept is "old news" with respect to OSHA's enforcement strategies. First, every targeting system that can be devised for the agency creates categories of workplaces that are "off limits" for regular inspections because the probability of injury or illness is relatively low. In no workplace, however, is risk entirely absent. But resource constraints compel OSHA—and other enforcement agencies—to adopt de minimis decision rules so that they will not be paralyzed by overwhelming case loads.

OSHA targets inspections: In effect, it identifies high-risk and de minimis risk categories of employers—on the basis of several types of information—and applies differing enforcement treatment to them accordingly.

First, there are industry injury rates, calculated annually by the Bureau of Labor Statistics. These rates are a powerful driving force for the safety inspection targeting system—they define high-risk industries where inspections will be concentrated.

Second, OSHA uses individual firm inspection histories to determine where problems have occurred previously—and, more important for this discussion, where they have not occurred. If a workplace is free of serious violations of OSHA standards when it is inspected, it is removed from the inspection targeting list for up to three years.

Third, OSHA looks at individual firm accident experience and compares it to industry averages as part of the normal safety inspection procedure. If it is sufficiently lower than industry experience, the inspection is ended with a check of the records. In these ways,

OSHA determines that the risks presented in a particular workplace, or in an industry, are to be considered "de minimis" and that enforcement resources should be placed elsewhere.

Next, OSHA operates two major exemption programs for firms which show that they have reduced risks to worker safety and health. These are the voluntary protection programs (STAR and TRY) and the consultation exemption program.

The consultation exemption program provides a one-year exemption from regular inspections for those small firms that (1) request assistance from OSHA-sponsored consultation programs, and (2) are operating an effective risk management program to protect employees from workplace hazards.

The VPP involve a (potentially) longer exemption and are focused primarily on the larger firms. These programs, too, require that an effective safety and health program be in operation in the workplace, and that workplace injury and illness records, and workplace exposure monitoring records indicate superior control of hazards.

At this point, having described a practical application of the de minimis principle in workplace targeting, it is appropriate to ask a question about its ethical aspects: Given the presence of risks to workers in every workplace, is it acceptable to exempt some workplaces on the basis of probabilities of risk and reviews of documents?

OSHA has answered that question with a qualified "yes." The major qualification is this: It is ethical *if* there is a safeguard which is sufficient to identify and to correct hazardous situations when they occur. In practice, this means using enforcement to compel correction of recognized but unabated hazards. In OSHA's case, a strong safeguard of this kind is provided by the worker's right to file a complaint with OSHA and to obtain an inspection and timely enforcement. None of OSHA's targeting mechanisms or of its exemption programs affect that basic right. Even workers in the most highly rated voluntary protection program may obtain an OSHA inspection if they believe that uncorrected hazards exist in their workplace.

Thus, in practical and ethical terms, de minimis already works for OSHA and is indeed necessary to operate a rational, efficient inspection program.

The question that follows, then, is "Can a practice that has operated well in the enforcement realm be applied to *standard setting?*"

Some significant parallels do exist between the standard-setting problem and the enforcement problem.

First, there is the overwhelming number of potential targets or candidates for regulatory action. In enforcement, it is a universe of some 5 million workplaces, more than 3 million of them in construction and manufacturing enterprises, where the most injuries occur. In the case of standards, it is an expanding universe of no fewer than 50,000 chemicals, and an uncounted number of mixtures of chemicals. And in both cases, only a relative few of the potential targets will develop into major threats to health and well-being.

Second, there is the similarity in OSHA's capacity to respond. In standards, as in enforcement, OSHA's capacity to respond will always be limited to a small fraction of the total universe of candidates, be they workplaces or chemicals. And in either case, even vastly increased resources would cover only a portion of the potential target areas. Further, in both cases, at some point, there is a decreasing marginal return on investment of new resources—a tenfold increase in staff and funds would not reduce hazards proportionally because those resources would be applied to progressively less hazardous chemicals and workplaces.

Third, in standard-setting as in enforcement, any prioritization system will leave some "bad actors" unregulated or underregulated, at least until known "worst cases" are resolved.

Given these parallels, however, one important difference stands out between enforcement targeting and standards prioritization; that difference is in the nature of the safeguard provided to override the policy judgments made on the basis of abstract assessments of risk.

In the enforcement targeting case, the right to obtain an inspection through filing a complaint protects employees from the rigidities inherent in the decision making process. In the standard-setting area, however, the choices are more clearly binary—either a standard for a given substance exists or it does not, and until the standard is on the books, it protects no one.

If the question is further limited to setting priorities for standards promulgation, the same rigidity again appears: Either the substance is on the list of priorities for standards development or it is not. If the prioritization process, however carefully constructed, passes it by, no OSHA standard will be issued for exposure to the substance. Several approaches can be considered to meet this need.

The first approach is to develop a successful prioritization system that is flexible enough to include decisions made by health professionals and administrators before all the necessary quantitative data are in: There must be room for professional judgment about what is likely to be a significant risk and what is not.

Second, alternative regulatory mechanisms must be available. OSHA has several features that may already fill this requirement. For example, section 5(a) (1) of the OSHAct requires that employers provide generally safe and healthful working conditions, free from recognized hazards. Thus, regardless of whether or not OSHA has prescribed an exposure limit for a particular substance, the recognition of that substance by the professional health community as hazardous may enable OSHA to require that exposures be minimized or eliminated.

Another, more specifically health-oriented safeguard is also available—that is the OSHA Hazard Communication Standard (29 CFR 1910.1200), which will increase the amount of information available to employers and employees about the known health effects of the substances they work with, in manufacturing industries. Under that standard, employers will be required to provide information and adequate training for a long list of substances over and above those specifically regulated by OSHA. That list includes all carcinogens and suspected carcinogens identified by IARC, and by the National Toxicology Program, and also other toxic substances, including those identified by the American Conference of Governmental Industrial Hygienists.

In that context, OSHA is exploring options for priority-setting systems. Two initial studies have already been completed, dealing with approaches to the prioritization process. We can also draw upon a wealth of experience with the analogous problem at other federal agencies, notably EPA. We will be reviewing all of this and asking the public—employers, employees, and health professionals—for their views on a prioritization system.

Given OSHA's practical acceptance and implementation of de minimis concepts, it is somewhat ironic that de minimis risk assessment might not be mentioned explicitly in official positions and policy statements. "De minimis" is, unfortunately, a term that is too easily misunderstood; it can be perceived as a way of "trivializing" a small but significant risk to health or safety. Nonetheless, de minimis concepts are inherent in every scheme for prioritizing standards candidates, and for allocating enforcement activity.

Finally, books such as this one make an important contribution to our understanding of the techniques of risk assessment. Certainly, OSHA will welcome the views of the whole scientific community, and particularly the views of those who have devoted time and professional attention to the problems of risk assessment, as we develop, test, and implement an improved priority-setting system.

10

Applications of De Minimis

Sheldon Meyers

The concepts of a nonthreshold pollutant (such as radiation) and of de minimis are essentially contradictory—the first states that there is no level of exposure that does not produce some harm, the second that there is some defined threshold below which there is no concern. It is the job of the regulatory agencies to see if these concepts can be reconciled and, more important, produce some benefit. As we shall see, this is a process of compromise and not of absolutes.

The derivation of de minimis as an abbreviation of the Latin phrase *de minimis non curat lex,* roughly translated as "the law does not concern itself with trivialities," denotes both its origin and intended purpose. "De minimis" is a term of convenience—it is intended for our purposes to be used in a regulatory framework to focus priorities. That is a very simple thing to do in theory, and very difficult in practice.

There has to be a reason to warrant consideration of a new concept in the regulatory arena. In this case, the incentives are both practical and economic. One naturally assumes that there should be a limit to the regulation of every risk in the universe, but it is most appropriate that such limits be based on rational and well-accepted principles. This may well be a utopian dream, as we shall see in a minute, and depends more on a perception of risk than on the actual magnitude of a risk. Nonetheless, if we could even vaguely define a de minimis level then this would constitute a very significant step in allocating resources for further risk reduction.

There are two possibilities for deciding that one eventually reaches a point where further risk reduction is not warranted: Either (1) the cost of further risk reduction becomes very great in relation to the small additional increment of benefit, or (2) the risk to the most exposed individuals and the entire affected population is so small that it becomes inconsequential and is of no concern to the great majority of people when compared to all other competing risks. Unfortunately, determination of a level of risk that is sufficiently

Sheldon Meyers • Office of Radiation Programs, Office of Air and Radiation, Environmental Protection Agency, Washington, D.C. 20460.

low so that people will routinely accept it as "of no consequence" is difficult because individual perception of risk differs greatly. Thus, while there may not be any single risk limit that will be acceptable to everybody in the population, a societal judgment can often be made that will satisfy most people.

Let us try to place de minimis in perspective. The Office of Radiation Programs recently sponsored a contract that provided a report entitled "A Measurement of De Minimis Risk." The authors tried to characterize a number of low-probability accident and health risks in quantitative terms of frequency of occurrence and assigned a qualitative measure of societal interest in terms of level of government activity to each. They concluded that there was no discernable lower limit of interest, even for events with a population-averaged lifetime probability of death of less than 1 per 10 million, or about 10^{-7} per year. Although both the assumptions and the conclusions of this report may be flawed, it nonetheless indicates that there may be no *a priori* way to establish a de minimis level and that the expenditure of resources for control of a risk may bear little relationship to the magnitude of the risk averted.

There are certain forces at work, however, that may force society to make some hard choices—and these forces are economic. With unlimited resources, the sole constraint is technology. With limited resources, the focus is shifted to a balancing of benefits and costs. We all know that each decade of risk reduction has generally increased costs and decreased benefits—it frequently is relatively cheap to reduce risks from 0 to 90%, more expensive to go from 90 to 99%, and more expensive still to go from 99 to 99.9%. It is also becoming increasingly important to compare risks across the entire spectrum and not for a single activity. Therefore, the impetus for risk management decisions that include consideration of de minimis is at hand.

The idea of publishing a statement on de minimis is not new—we considered such a move more than half a decade ago and decided to wait because the time was not yet ripe. Regulatory agencies generally do not lead but tend to seek a consensus position. We are now closer to the point when this idea can be accepted. What will be important is to carefully consider all the pros and cons and to do our deliberations in full public view.

Such vague definitions of de minimis as "harmless," "innocuous," or "negligible" quantities are not very useful from a regulatory perspective. De minimis is really an administrative concept—it deals with a societal judgment of what is acceptable for certain defined situations. Such a universal concept is almost impossible to quantify on an overall regulatory basis. Therefore, we at EPA prefer not to use the generic term but rather to use the term "below regulatory concern" (BRC) and limit it only to certain narrowly defined specific objectives. There may then be a series of BRC levels, depending on feasibility, costs, and other considerations.

The concept of establishing levels of BRC is most attractive, especially in those areas where the concentrations of materials that may be released to the environment are very low and the costs of alternative reduction or disposal methods is rather high. The most obvious such application is to certain categories of low-level radioactive waste. That has been the only application to date and is obviously important from an economic standpoint.

There are two incentives for developing a concept of de minimis (or of BRC) for disposal of low-level radioactive waste. The first is the saving in costs for disposal—all the extra costs of special handling, packaging, transportation, and disposal. The second is the saving in land—the reduction in the number and size of disposal facilities that may be needed. The magnitude of the savings is a function of where one places the BRC level. The higher the exemption level, the greater the potential savings.

Several determinations of "exempt quantities," quantities of radioactive materials that may be disposed of without further restrictions, have already been made by the Nuclear Regulatory Commission. These include the quantity of americium-241 in a smoke detector, and defined limits of tritium and carbon-14 in liquid scintillation vials and biological waste. The first decision was made purely on the basis of feasibility—it simply is not practical to maintain control of smoke detectors at the consumer level. The second decision established a process that considers both public health protection and costs of alternative disposal methods. Thus, the process is already under way, but purely on an *ad hoc* basis and without any general guidelines. We believe that the time has come for such guidelines to be properly debated and codified.

The Task Force on Low-Level Radioactive Waste of the former U.S. Radiation Policy Council stated in 1980 that "an overall generic de minimis level is not a practical solution to a portion of the waste management problem because, without knowing physical and chemical parameters of the waste involved, it is difficult to establish pathways to humans and resulting doses. Therefore, in order to establish a generic de minimis level, extremely conservative assumptions are dictated which are likely to lead to levels that are so small that they have little practical value in disposing of low activity waste. As an alternative, the Task Force endorses the approach of evaluating waste streams on a case-by-case basis. . . ." It is this approach that EPA has adopted and we are now developing the scientific basis for such a determination.

Let me digress for a moment and discuss some of the alternatives. A rationale based on risk alone is most attractive; however, we cannot justify a reason for limiting such a cutoff to radiation alone, and it would be almost impossible to define a universally agreed level for all hazardous materials. A rationale based on some small fraction of background radiation levels has also been proposed. This leads to the questions of why and where: Why increase an existing risk, and where is the appropriate stopping place? Therefore, neither the absolute value of the background radiation nor its variability from place to place gives a clear basis for selecting a level of de minimis or BRC. A third rationale is that ALARA leads directly to a de minimis level. However, ALARA is implemented to reduce a regulatory limit and works from the top down rather than the reverse. Thus, the two concepts operate on a different basis, and there is no distinct single meeting point between ALARA and de minimis. The application of one does not necessarily lead to the other.

Our analyses of BRC for low-level waste have focused on an evaluation of the consequences of disposal of the basic waste stream categories by a number of the proposed disposal methods, and they are continuing. We published an advance notice of proposed rule making last fall to develop criteria for BRC, and we hope to propose numerical criteria by next spring. The development of such regulatory criteria for practical application to low-level waste disposal is obviously in an evolutionary period. We are certain that our proposal will generate much interest, and will carefully consider all comments we will receive. It is by such a process that allows input from both the scientific community and the general public that a consensus on levels of "below regulatory concern" can evolve. EPA will take a lead role in that process.

Let me also address the question of acceptance. We are all aware of the mystique of radiation. Public perception is such that even miniscule risks are often viewed out of context. Therefore, there is no assurance that, even if a regulatory agency declares certain categories or quantities of radioactive materials as de minimis or below regulatory concern, the public will accept such a declaration. We could face a quandary such as a landfill operator or a municipality refusing to accept such wastes even in the face of a regulatory judgment. A determination of BRC may well need to be coupled to a public education program that seeks to educate the general public on the risks from radiation.

In conclusion, let me make several observations. The first is that de minimis risks cannot be considered in total isolation because all environmental risks are essentially additive. Thus, while a single risk increment of perhaps one per million per year would generally be considered to be very small, people are exposed to a very large number of carcinogens—each with its own probability of causing some harm.

The second is that the size and characteristics of the population used in the analyses may be of importance. A defined de minimis risk limit applied to a critical segment of the population is fundamentally different from one applied uniformly to the entire population. In the first case, only a few people would be expected to be at a maximum risk, and the average is probably one or two orders of magnitude lower. This is the situation applicable to most local environmental contaminants where specific sources can be identified, as is the case for most radiation exposures. In the second case, a much larger segment of the population would be at maximum risk, and the average risk may not be much lower. This is the situation applicable for widely dispersed pollutants, such as automobile emissions or certain food additives. In general, when the risk is already very low, society appears to recognize a difference between these situations, and to accept a somewhat higher risk limit when the number of exposed individuals is small. The rationale for such a difference in risk perception may well be an implicit balancing by the public of costs of risk aversion per individual against the resultant benefits to society.

The third is that we need to distinguish between de minimis and zero risks. "Zero" is a relative rather than an absolute term and can only signify the level below which no measurable concentration or effect can be detected. Such a zero risk could well be outmoded at a future date, the cost of achieving it is likely to be prohibitive, and it totally precludes any regulatory judgments.

Finally, we must also remember that the components of mixed wastes, which may contain more than one hazardous material, cannot be considered in total isolation. Regardless of whether the disposal of radioactive wastes would result in less then de minimis consequences, the nonradioactive (chemical and/or biological) hazards of the waste must also be considered, and disposal must be in accordance with regulations applicable to the most hazardous component. Therefore, de minimis quantities of wastes from the radiological viewpoint may still be subject to restrictions if other types of hazards are present.

We have seen that the subject of de minimis is very complex, and no easy answers are at hand. Nevertheless, I think that it is most useful to explore the possibilities and determine specific applications on a limited scale. It seems to me that the benefits of applying a de minimis rationale to the disposal of certain wastes are quite evident and that rational judgment can be made by the regulatory agencies. We solicit the help of the technical community in developing an appropriate basis for our decisions.

11

The NCRP Considerations on Levels of Negligible Risk

Charles B. Meinhold

NCRP's Committee 1, under the chairmanship of George Casarett, has approached the de minimis question from the standpoint of risk. In the document approved by the council there is a recognition that even with the prudent assumption of a linear relationship between risk and dose, there must exist a risk and its attendant dose that is negligible (NCRP 1987).

Of primary importance in this issue of developing a de minimis level is the use of risk as the fundamental quantity of importance in deriving such radiation protection recommendations.

An in-depth review of the known literature for estimating risk from radiation exposure (particularly low-LET, low-dose and dose-rate) is an inherent part of the Committee 1 efforts. As one might expect, the risk associated with whole-body exposure, used in deriving a de minimis quantity, is not significantly different from that of UNSCEAR, ICRP, and BEIR. This, together with the development of risk comparisons, allows a judgment of negligible radiation risk to be analyzed and presented in the context of more commonly understood risks of everyday life.

It is essential that a model adopted for risk estimation have a strong scientific basis. As with UNSCEAR, ICRP, and the majority of the BEIR committee, NCRP Committee 1 found the most reasonable relationship to be one that incorporates a linear term that predominates at low dose, and a dose-squared term that predominates at high doses. It may well be that much of the difficulty in arriving at a consensus on a de minimis dose lies in the commonly held belief that these UNSCEAR, ICRP, and BEIR risk estimates grossly overestimate the actual expected increased incidence of cancer in a population exposed to low doses.

Charles B. Meinhold • Safety and Environmental Protection Division, Brookhaven National Laboratory, Upton, New York 11973.

Although there are no conclusive data to prove that there is risk at low doses, the available data do suggest that linearity at low doses is the relationship one would expect from first principles—both physical and biological. As can be seen in NCRP 64, "Influence of Dose and Its Distribution in Time on Dose–Response Relationships for Low-LET Radiations," simple biological systems clearly demonstrate that even with low-LET radiation there is a linear component throughout the low-dose range. The data from most animal and human studies do not contradict this assumption.

It is the threshold or quasi-threshold relationship that, however plausible, must be proven. NCRP Report 64 concluded, however, that the slope of the linear portion will vary with the biological system under study. The risk estimates currently in use include a dose rate effectiveness factor (DREF) and as such are not based on a simple extrapolation of effects seen with absorbed doses at 100 to 200 rads. It was because UNSCEAR-77 in effect used a dose rate effectiveness factor of 2.5 that ICRP noted in their 1978 Stockholm meeting statement that "these risk factors are intended to be realistic estimates of the effects of irradiation at low annual dose equivalents."

Both NCRP and ICRP have recommended a nonoccupational dose limit of 100 mrem per year for individual members of the public based on a realization that the risks to which the public is exposed range from 10^{-2} per year to 10^{-6} per year, and that annual risks of 10^{-5} to 10^{-6} are associated with those activities considered to be accepted freely by most people. The NCRP, therefore, suggests that an annual risk limit of 10^{-5} and an annual dose equivalent limit of 100 mrem (1 mSv) be applied for continuous or repeated exposure of individual members of the public from all sources with the exception of medical and natural background exposure. It is noted that this risk level is not significantly different from the average risk from natural background or that from medical exposure.

With regard to this limit on nonoccupational exposure, there is the associated requirement for justification and ALARA just as there is with the "surrogate" risk actively used for developing these comparative risk values. Reasonable efforts to reduce accidents in our "safe" activities are expected and do take place. For example, the risk level in "safe" industries is decreasing at a significant rate as a result of the application of justification and ALARA. We are therefore constrained to reduce the nonoccupational risk below 10^{-5}, but should we be constrained to reduce this individual risk to 10^{-6}, 10^{-10}, or 10^{-12}? Most reasonable people would agree that there must exist an insignificant risk rate.

Just how far below this limit we should attempt further reduction is "stuff" of which de minimis dose is made. Committee 1 believes that such a level of risk exists and that it might be appropriately called a reasonably negligible risk.

The societal benefits to be gained from the application of the reasonably negligible risk level in the radiation protection field, and as well in other fields of health protection, include (1) aid in establishing rational criteria for shielding design, sensitivity of environmental monitoring instruments, recording of individual exposures, and other activities in health protection; as well as (2) education and assurance of a concerned public as to the distinction between levels of risk, i.e., significant versus trivial. The negligible risk level could produce economic benefits in many areas, e.g., avoidance of expensive, futile epidemiological studies at very low exposure levels, an elimination of many analyses done to estimate environmental impact, or cost–benefit balances where exposures are calculated to minute levels and integrated over vast expanses of space and time. Time and efforts of scientists, technologists, supporting staff, and regulatory personnel, as well as funds and other societal resources, would then be available for study and control of the nonnegligible risks associated with environmental agents or conditions, with potential saving of many more lives.

It is true that exposures of any magnitude may be calculated, even if not measurable. In monitoring, exposure planning, or assessment, the annual radiogenic risks (based on the health effects criteria in this report) that are lower than the negligible risk level (NRL) would not be included, in either individual or collective assessments. Utilizing a negligible risk level establishes a threshold below which control of radiation sources, practices, and exposures—i.e., efforts to reduce risk further—would be deliberately and specifically curtailed.

The NRL is here regarded as a lower limit to ALARA. However, it is not, *per se*, the goal of ALARA, which in principle and by definition should be determined by what is "reasonably achievable" below the recommended maximum permissible limits. ALARA requires every reasonable effort to maintain exposure to radiation as far below the maximum limits as is practicable, taking into account the state of technology, the economics involved in relation to public health and safety benefits, and other societal and socio-economic considerations.

Of primary importance is the recognition and acceptance of the concept of reasonable negligible risk level(s). These levels can be adjusted to meet new information or changing criteria. Even if the NRL were set initially at a very low value, considerable benefit may accrue in terms of elimination of concern that currently exist about extremely small doses. It is also of *great* importance that the NRL be consciously low, as intended.

Furthermore, it is important in the setting of the NRL to use approaches that would minimize subjective judgments to the degree possible, and that would tend to satisfy the requirements for public understanding. Committee 1 considered several approaches that, taken together, offer degrees of reasonability and perspective that go far toward meeting these requirements. These approaches or criteria are concerned with (a) smallness of risk in relation to magnitude of dose; (b) difficulty in measurement (of dose or risk); (c) observed natural risk for the same health effects (e.g., cancer); (d) risk estimated for the mean or the variance of natural background radiation exposure levels; (e) accustomed risk levels (e.g., in ordinary, normal living activities); and (f) perception of risk (behavioral response).

Smallness of risk in relation to magnitude of dose is a factor since the radiation health risk decreases with decrease in dose, becoming very low at very low exposure or dose levels.

Smallness of risk in relation to difficulty in measurement of either effect or dose is considered since it is highly unlikely that the assumed health risks from radiation exposures of the order of those received from natural background sources can be measured directly and unequivocally.

The difficulties in measurement at very low radiation exposures pertain not only to epidemiological studies but also to the physical measurement of increments of exposure against background.

Smallness of risk in relation to natural risk for the same health effects was considered. This comparison of the health risks estimated to be associated with low-level radiation exposure with the relatively high natural risks of the same effects from all causes provides useful perspective and illustrates one of the major reasons for the difficulty in measuring the radiogenic health risks at low exposure levels. For example, the risk (from all causes) of developing cancer among members of the U.S. population is roughly 1 chance in 4, or 25%, and the risk of developing fatal cancer is roughly 15%. The lifetime risk rate for developing radiation induced fatal cancer is estimated to be roughly 1 in 10,000 per rad or 1 in 100 per Gray.

Smallness of risk in relation to estimated risk for mean or variance of natural background radiation exposure levels was considered. Although there are reasons for assuming that natural background radiation exposure may carry risks of cancer and hereditary effects, such risks are expected to be so small in comparison with the natural risk as to be beyond direct unequivocal ascertainment.

The dose equivalent received from natural background radiation varies by a factor of about 3 in the regions inhabited by substantial fractions of the world's population, and the mean dose equivalent to the population is roughly 0.1 rem (1 mSv) per year, excluding the lung dose due to radon. The standard deviation in the United States from low-LET radiation is about 25 millirad per year (0.25 mSv per year), with a mean of about 100 millirad per year (1 mGy per year). A value of 20 millirad per year (0.2 mSv per year) has been suggested as the definition of "small" compared to natural background (Adler and Weinberg 1978).

Smallness of risk in relation to accustomed risks was examined because the comparison of estimated radiogenic risks with other risks involved in a variety of activities of living offers valuable perspective.

A fatality risk of about 10^{-5} per year is close to the level below which there is little concern—i.e., an activity is regarded as reasonably safe. The draft suggests that a risk level two orders of magnitude lower, i.e., 10^{-7} per year, would appear to be trivial and negligible.

Smallness of risk in relation to perception (behavioral response) is important since the concept of risk as finite probability is alien to many members of the general public, who tend to prefer a simple answer to the question of whether something is safe or not safe. Furthermore, the assessment of safety is not necessarily based on probabilities alone. The question of negligibility requires resolution before numerical values are attached. Behaviorally, a situation may be perceived or regarded as safe—i.e., the associated risks negligible—if reasonably informed persons, when offered a safer alternative, ignore it in favor of perceived benefits.

It appears that society has not been significantly concerned about differences in dose due to variations in external natural background radiation or due to different natural building materials. However, it is possible that lack of knowledge plays a large part in this.

In view of the preceding considerations annual risk commitment increments of 10^{-7} or less, or total lifetime risks between 10^{-5} and 10^{-6} or less, are regarded as reasonably negligible.

The report recommends that application of the negligible levels of risk or dose to the individual would also, in application of the individual NRL in collective dose determinations, limit the collective dose appropriately.

The absorbed dose that would commit an individual to a negligible risk level of 10^{-7} yr is 1 mrad (10 Gy) of low-LET radiation per year. This is 1% of the dose limit recommended for members of the public.

Up to this point two recommendations regarding dose limitation for members of the public have been discussed: first, a dose equivalent limit of 100 mrem per year (1m Sv per year) based on an acceptable annual risk of 10^{-5} and a negligible risk level of 1 mrem per year (10 μ Sv per year) based on a negligible annual risk of 10^{-7}. It should be clear, however, that there are other levels that exist between these two based on apportionment, justification, and ALARA.

For example, the NCRP Statement on the Control of Air Emissions of Radionuclides, September 18, 1984, suggested an action level of 25 mrem per year (0.25 mSv per year)

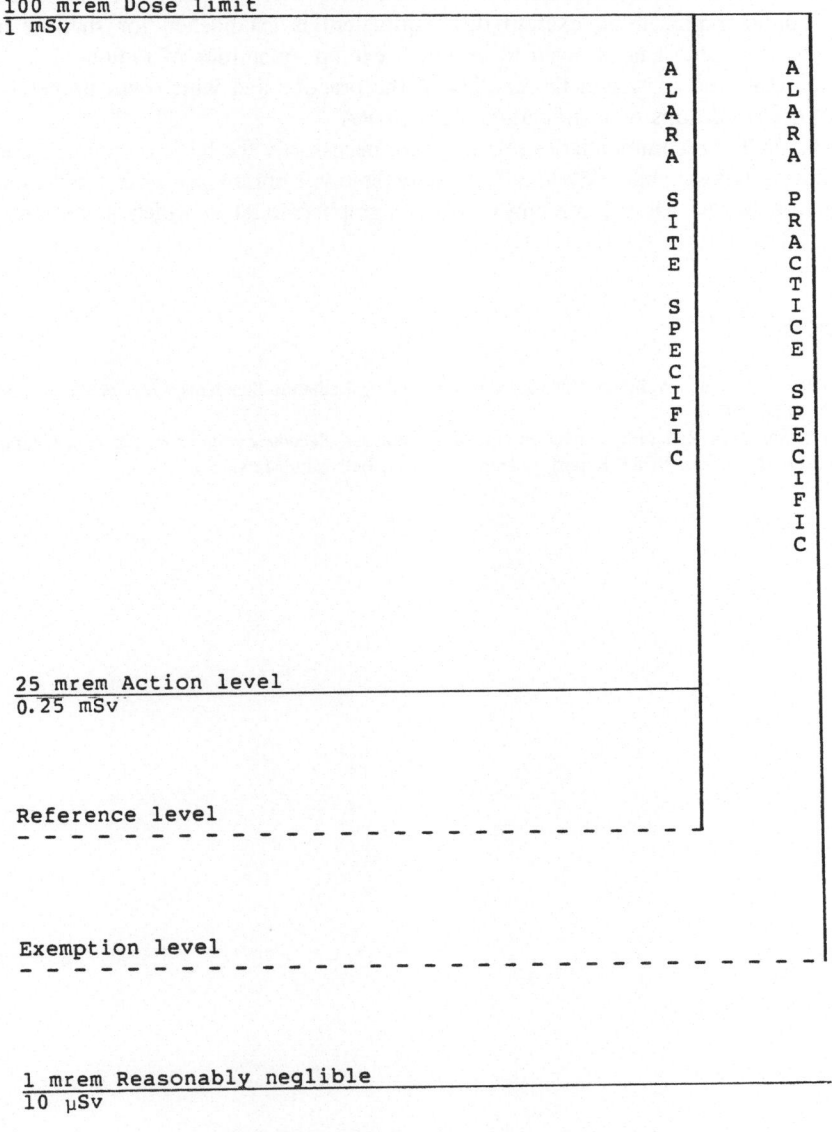

Figure 1. Nonoccupational annual limits and levels.

to the whole body to ensure that discharges from multiple sources would be unlikely to result in an individual's exceeding the 100 mrem-per-year (1 mSv-per-year) limit.

With regard to individual sites or practices, there might well be similar considerations. This approach, however, leads not to a negligible risk level but rather to an authorized limit or action level. In the application of the limit it is still necessary to apply ALARA considerations. Such considerations may lead to a reference level if they are derived on the basis of a specific site or practice. If, however, the justification and ALARA are generic in nature, the dose level may well be the basis for an exempt or "below regulatory interest" level. This might be diagrammed as illustrated in Figure 1.

Any of these dose-equivalent levels will have derived quantities associated with them. For example, if an exempt dose-equivalent is established for shallow landfill disposal, this value can be used to establish exempt quantities of radionuclides which could be disposed of within the context of the practice and which was properly tested against the conditions of justification and ALARA.

The NCRP recommendation, then, is not necessarily the basis of exempt quantities but rather suggests that individual exposure below 1 mrem per year requires neither justification nor ALARA. It remains with lexicographers to tell us which of these quantities is de minimis.

REFERENCES

H. L. Adler and A. M. Weinberg, "An Approach to Setting Radiation Standards," *Health Physics, 34*, 710–720, (June 1978).

National Council on Radiation Protection and Measurements, *Recommendations on Limits for Exposure to Ionizing Radiation*, NCRP Report 91 (June 1, 1987), Bethesda, Maryland.

12

A Summary Perspective on NRC's Implicit and Explicit Use of De Minimis Risk Concepts in Regulating for Radiological Protection in the Nuclear Fuel Cycle

Miller B. Spangler

TOWARD A COMMON-USE CONCEPT OF DE MINIMIS RISK

There is a growing recognition of the desirability in public policy making regarding a cutoff standard of insignificant risk to base this on reference values of common use that roughly characterize the levels of individual risk that are regarded as sufficiently negligible not to merit additional personal expenditures to reduce them further. According to Clarke of the U.K. National Radiological Protection Board, there is a widely held view that few people would commit their own resources to reduce an annual risk of death of 1 chance in 100,000 and that even fewer would take action at a chance of 1 in 1 million per year of exposure to a given hazard.[1]

It is to be noted that there is an important distinction between "acceptable risk" and an "acceptable standard of de minimis risk," upon which this chapter focuses. All risk levels regarded as insignificant (or de minimis) should, by this very same token, be regarded as acceptable risks. However, society finds certain other nontrifling risks to be acceptable whenever the expected *net* beneficial effects in a risk-cost-benefit (RCB) analysis are perceived to outweigh the expected adverse effects associated with individual or public hazards. This principle is implicit in the concepts presented in Figure 1 as reproduced from an article by Richard Wilson.[2] The numbers in the left-hand column on risk of individual death were derived from the work of Lord Brian Flowers, chairman of

Miller B. Spangler • Office of Nuclear Reactor Regulation, Nuclear Regulatory Commission, Washington, D.C. 20555.

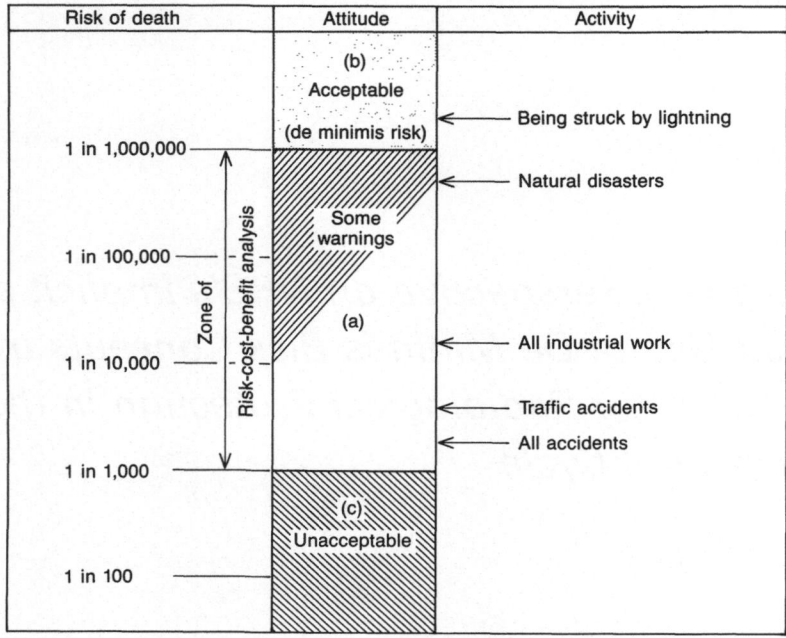

Figure 1. Probability of death for an individual per year of exposure (orders of magnitude) in terms of acceptable/unacceptable risk.

the Royal Commission on the Environment of the United Kingdom. The use of an inverted scale of risk reflects the desirability of expressing an increasingly higher standard of safety. Accordingly, the upper zone of Figure 1, with risk of individual fatality per year of exposure that is less than 1 chance in 1 million (i.e., 10^{-6} per year), might appropriately be regarded as an acceptable standard of de minimis risk. On the other hand, the lower zone, with individual risk greater than 1 chance in 1000 per year of exposure, might be regarded as clearly unacceptable. The middle, in-between zone would then be the valid arena of debate for risk-cost-benefit analyses to be performed in social decision making in determining whether risks are acceptable or not. According to Clarke,[1] most authors proposing individual dose cutoffs for radiation protection have set the level of annual radiation risk that is held to be of no concern to the individual at 10^{-6} to 10^{-7}.[3,4,5]

Guidance for the use of value judgments in a determination of acceptable risk in radiation protection is provided in ICRP Report No. 26, which recommends a system of "individual-related" protection with dose limitations based on the following principles.[6]

- No practice shall be adopted unless its introduction produces a positive net benefit ("justification of practice").
- All exposures shall be kept as low as reasonably achievable (i.e., ALARA), economic and social factors being taken into account ("optimization of protection").
- The radiation doses to individuals shall not exceed the dose equivalent limits recommended for the appropriate circumstances by ICRP.

A different safety philosophy is "source-related" protection, which is based on the principle that optimization should include assessment of how much the total radiation from a given source would decrease owing to various possible protective measures. Even if the risk of harm to each exposed individual were very small, exposure of a large number of individuals might cause a collective harm that could perhaps be avoided at reasonable cost and effort. As noted by Beninson and Lindell,[7] the management of an installation that releases radioactive substances into the atmosphere may claim that the individual doses at some distance from the release point are "de minimis" and may therefore be neglected. Since these small doses may cause a sizable collective dose that might be reduced at reasonable effort, the concept of a societally aggregated, or source-based, de minimis risk standard is also desirable to consider according to this view. Both an individual-based approach and a source-based (or societally aggregate) approach are to be found in NRC regulatory philosophies and are included in the discussions here.

NRC REGULATION OF ACCEPTABLE (NO UNDUE RISK) LEVELS OF RADIOLOGICAL PROTECTION IN NORMAL ACTIVITIES OF THE NUCLEAR FUEL CYCLE

The various processes of the nuclear fuel cycle include mining and milling, fuel processing and enrichment, power generation, the transport of fuels and wastes, reprocessing of spent fuels (when permitted), and waste management. By "normal" activities is meant those activities and procedures of the fuel cycle not involving severe accident policy issues, which will be addressed in the next section. Thus, normal activities covered in this section include design basis events and severe accident prevention measures associated with the routine operation and maintenance of nuclear power plants as well as other routine activities of the nuclear fuel cycle where de minimis risk concepts might apply.

The ALARA Concept of Radiological Protection

The NRC and its professional staff participate in or take cognizance of the deliberations and reports of the International Commission on Radiological Protection (ICRP) and a number of other organizations and meetings relating to radiological protection. This also includes coordination with the standard-setting radiological protection activities of the U.S. Environmental Protection Agency. However, that an accounting of these interrelationships is beyond the scope of this chapter should not be taken as an underreflection of their considerable importance to NRC's historical development of acceptable risk policies, standard-setting rules, regulatory guidelines, technical specifications, or branch technical positions. Indeed, the need for brevity permits only a limited coverage of the more salient NRC efforts relating to NRC's implicit and explicit use of de minimis risk concepts.

The Energy Reorganization Action of 1974 and the Atomic Energy Act of 1954 have provided only qualitative guidance on which the NRC must base its policies and rules regarding its primary responsibilities for protecting the public health and safety. For example, Section 103b of the Atomic Energy Act of 1954 on the subject of commercial facility licenses states:

b. The Commission shall issue such licenses on a non-exclusive basis to persons applying therefore (1) whose proposed activities will serve a useful purpose proportionate to the quantities of special nuclear material or source material to be utilized, (2) who are equipped to observe and who agree to observe such safety standards to protect health and to minimize danger to life or property as the Commission may by rule establish, and (3) who agree to make available to the Commission such technical information and data concerning activities under such licenses as the Commission may determine necessary to promote the common defense and security and to protect the health and safety of the public. All such information may be used by the Commission only for the purposes of the common defense and security and to protect the health and safety of the public.

Also, Section 161b of the Act gives the Commission authority for standard setting. Additional relevant provisions of the Energy Reorganization Act of 1974 regarding risk-cost-benefit analysis and related NRC regulatory functions are as follows:

- Title III, Section 307(c) requires NRC to make a clear statement annually of the short-range and long-range goals, priorities, and plans of the Commission as they relate to the benefits, costs, and risks of commercial nuclear power.
- The kinds of "benefits" the Congress had in mind in passing this Act are found in Section 2(a) under "Declaration of Purpose:"

Section 2(a). The Congress hereby declares that the general welfare and the common defense and security require effective action to develop, and increase the efficiency and reliability of use of all energy sources to meet the needs of present and future generations, to increase the productivity of the national economy and strengthen its position in regard to international trade, to make the Nation self-sufficient in energy, to advance the goals of restoring, protecting, and enhancing environmental quality, and to assure public health and safety.

However, in the protection of public health and safety, many of NRC's regulations as established in Title 10 of the Code of Federal Regulations have been principally guided by the subjective criterion of "no undue risk" to public health and safety:

(10 CFR Part 50, Appendix M, Section 5d) On the basis of the foregoing safety review considerations for nuclear power reactors of standard design, the Commission may issue a license if there is reasonable assurance that (i) such safety questions will be satisfactorily resolved before any of the proposed nuclear power reactor(s) are removed from the manufacturing site and (ii) taking into consideration the site criteria contained in Part 100 of this chapter, the proposed reactor(s) can be constructed and operated at sites having characteristics that fall within the site parameters postulated for the design of the reactor(s) without undue risk* to the health and safety of the public.

(10 CFR Part 100, Section 100.10) Among other factors, the Commission in determining the acceptability of a site for a power or testing reactor will take into account Appendix A, Seismic and Geologic Siting Criteria for Nuclear Power Plants, which describes the nature of investigations required to obtain the geologic and seismic data necessary to determine site suitability and to provide reasonable assurance that a nuclear power plant can be constructed and operated at a proposed site without undue risk to the health and safety of the public.

A major step in NRC regulatory decision making involved rulemaking leading to the adoption of ALARA radiation protection standards for the design and operation of

---

* Regulatory decision making relative to the "minimization of danger to life or property" or the "avoidance of undue risk" implicitly involves cost considerations without which risk would be pressed to the zero level by unwarranted expenditures on behalf of safety (e.g., containment structures with 100-foot-thick walls, population exclusion zones of hundreds of square miles, dozens of standby diesel generators for auxiliary feedwater systems, etc.—to exaggerate the point). See also a 1979 memorandum by NRC's Office of General Counsel on the definition of "Adequate Protection of the Health and Safety of the Public."[8]

light water reactors (LWRs) in May 1975.[9] The purpose of this regulation is set forth in 10 CFR 20.1(c).

(c) In accordance with recommendations of the Federal Radiation Council, approved by the President, persons engaged in activities under licenses issued by the Nuclear Regulatory Commission pursuant to the Atomic Energy Act of 1954, as amended, and the Energy Reorganization Act of 1974 should in addition to complying with the requirements set forth in this part, make every reasonable effort to maintain radiation exposures, and releases of radioactive materials in effluents to unrestricted areas as low as is reasonably achievable. The term "as low as is reasonably achievable" means as low as is reasonably achievable taking into account the state of technology, and the economics of improvements in relation to benefits to the public health and safety, and other societal and socioeconomic considerations and in relation to the utilization of atomic energy in the public interest.

The ALARA concept, of course, is not synonymous with the de minimis risk concept since the risk-cost-benefit analysis used in establishing an acceptable level of risk may arrive at a risk standard that, in some cases, could be greater than an insignificant (or de minimis) level of risk. Yet, as illustrated in Figure 1, the use of ALARA-type procedures within the zone of risk-cost-benefit analysis could also provide a basis for establishing an explicit standard of de minimis risk beyond which no further analysis of costs and benefits need be employed to determine the acceptability of risk.

The regulation involved quantitative design objectives and an interim numerical safety–cost trade-off criterion as a guide to the desirability or undesirability of further expenditures to increase the level of radiological protection. The design objectives incorporated in the new rule (10 CFR 50, Appendix I) include the following:

1. Limit the amount of radioactivity released in liquid effluents from any light-water-cooled power reactor to levels that would keep the annual exposure to an individual in an unrestricted area to not more than 3 millirems for the whole body and not more than 10 millirems to any organ.
2. Limit releases of radioactivity in gaseous effluents from any light-water-cooled power reactor to keep annual exposures to an individual in an unrestricted area to a maximum of 5 millirems to the whole body, and not more than 15 millirems to the skin.
3. Limit releases of radioactive iodine and other radioactivity from any light-water-cooled power reactor to keep annual exposures to the thyroid of an individual in an unrestricted area to no more than 15 millirems.

During the rulemaking proceeding for Appendix I to 10 CFR Part 50, evidence was received relating to the establishment of useful criteria for deciding the appropriateness of reducing exposures to the population even further. Contributions to the record were in terms of costs of further measures compared to the benefits of any reduction achieved. These ranged from $10 to $980 for the whole exposure to a unit of collective (population) dose, the person-rem. (The person-rem is a measure of exposure to radiation of large groups of people—for example, 100 people each being exposed to 0.01 rem, 1000 people each being exposed to 0.001 rem.) In its decision, the Commission said:

The record, in our view, does not provide an adequate basis to choose a specific dollar value for the worth of decreasing the population dose. . . . We propose, therefore, at the earliest practicable date to conduct a rulemaking hearing to establish appropriate monetary values for the worth of reduction of radiation doses to the population. . . .

Meanwhile, and purely as an interim measure, we believe that we can accept the conservative value of $1000 per total body person-rem for these cost-benefit evaluations. Since we realize that the ultimately accepted value may well prove to be less than this, we should leave it open to demonstration in individual cases that a lower figure should be used if the applicant chooses to and can make that demonstration.

The value of $1000 per person-rem averted is roughly equal to $7.4 million per fatality averted using a ratio of 135 fatalities per million person-rems. The latter ratio employed in a recent Environmental Impact Statement (EIS) by the staff[10] is somewhat above the midpoint of the ranges of values for fatalities per million person-rems estimated by the 1980 BEIR-III Report (67–169) and the 1977 UNSCEAR Report (75–175). Thus, the ALARA value established *temporarily* by the commission is substantially higher than the equity* value of $250,000 to $500,000 per fatality averted referenced by other agencies in risk reduction decisions.[11] With reference to Figure 1, this equity value—in this case, conservatively chosen—provides the principal criterion along with estimates of other costs and benefits in deciding whether a risk, if left in Zone (A), is acceptable or not.

Regarding the subject of de minimis risk, if one uses the ratio of 135 fatalities per million person-rems, then the above upper risk bound in Appendix I of 5 mrem of whole body dose to the individual in an unrestricted area from gaseous effluents is computed at a probability of 6.7×10^{-7} fatality per year of exposure, or slightly less than 1 chance in 1 million for premature fatality. Likewise, a maximum exposure of 3 mrem from liquid effluents to an individual in an unrestricted area in the vicinity of a nuclear power plant is computed as yielding a probability of 4 chances in 10 million of premature fatality per year of exposure. In either event, the individual risk of fatality is within the range of common-use values suggested in the preceding section as qualifying as de minimis risk from hazards falling below these upper bounds. On the other hand, the standard limiting an individual in the general population to 500 mrems of annual exposure from all sources of radiation except natural background and medical sources (see above) would yield a probability of premature fatality of 6.7 chances in 100,000 per year of exposure. This value would not be regarded as de minimis according to most usages of that term for individual risk. In this event, it would be desirable to seek ways of mitigating the risk using a safety–cost trade-off value such as the $1000 person-rem in NRC's ALARA concept.

The NRC standards for occupational radiation dose in restricted areas are much less stringent than for individuals in areas of unrestricted public access. For example, in 10 CFR 20.101a the whole-body dose limit for occupational workers is a maximum of 1.25 rems in any one quarter and may be increased to 3 rems per quarter under conditions specified, or a total of 12 rems per year. This is 10 times the maximum for an individual in the general population and is also substantially above what might be regarded as a de minimis level, especially if such a dose were repeated year after year.

NRC's current regulations on permissible doses, levels, and concentrations of radionuclides were first published in Part 20 of 10 CFR in January 29, 1957 (22 FR 548).

---

* The term *equity value* is used for the safety–cost trade-off criterion to avoid the undesirable connotation of putting a price on human life, which, after all, is priceless. The issue of equity is at stake, however, since a policy that leads to greater expenditures per life saved than numerous alternative and unrealized opportunities for saving lives is an inequitable commitment of society's resources that otherwise could have been used to save a greater number of lives.

A proposed revision of Part 20 covering numerous aspects of NRC's "Standards for Protection Against Radiation" was forwarded to the Commission on April 22, 1985, for approval for issuance as a proposal for public comment.[12] One section of this proposed revision deals explicitly with establishing a policy of de minimis risk (pp. 8–9 of the Commission paper):

> De Minimis
>
> The need has long been recognized for a de minimis feature in the standards for protection against radiation in order to avoid extending regulatory actions beyond what is needed to adequately protect public health. However, the present Part 20 provides neither a de minimis level for the most exposed individual member of the public, nor a level for the cutoff of collective dose calculations. The lack of these levels in radiation protection standards has resulted in unwarranted expenditures of resources for incremental risks which are considered trifles in comparison to the risks which individuals are subjected to daily as part of normal living habits and activities.
>
> The Supplementary Information of the proposed revision of Part 20 discusses the possibility of a provision which would define a de minimis source of radiation as one which would be unlikely to cause any individual to receive a dose in excess of 0.001 rem in a year. . . . This concept of de minimis has received favorable endorsement from both the technical and legal standpoints. However, concerns have been raised that such a provision could be abused by licensees, in particular with respect to radioactive waste applications, and that it would permit uncontrolled use of radioactive material in consumer products. The revision introduces the concept of de minimis values, but proposes only a limited application of the concept by allowing licensees to use a cutoff value of 1 mrem (0.01mSv) in a year to individuals in calculations of collective dose.
>
> General acceptance of this concept would, however, make possible its broader consideration in addressing important regulatory questions, such as defining low-level radioactive waste. Comment on this issue is specifically requested in the notice of proposed rulemaking.

It should be noted that the proposed cutoff value for de minimis risk of 1 mrem in a year to individuals in the calculation of collective population dose is roughly equivalent to an excess fatality rate (i.e., premature death) of about 1.4 chances in 10 million per reactor year of operation. This is about a factor of 7 lower than the 10^{-6} per year level suggested in Figure 1. However, the establishment of one numerical cutoff level as a de minimis (or insignificant) risk for a certain class or classes of regulatory decisions does not, in principle, mean that some higher (or lower) level of risk might also come to be regarded as insignificant for other classes of decision. For example, the Commission has not yet addressed the question as to what is the *highest* risk number it would regard as de minimis for *any* class of regulatory decision involving regulatory protection. Nor has an integrated philosophy been developed as yet that would justify the use of different cutoff levels for different classes of regulatory decision and, indeed, whether an aggregate societal standard of de minimis risk would be preferable to an individual perspective of de minimis risk or vice versa.[13] Many considerations might affect such choices of philosophies and cutoff levels, including whether the risk is from routine emissions or catastrophic accidents, and whether the uncertainty surrounding the estimates of the risks is broad or not. Obviously, future advances in scientific knowledge that reduce such uncertainty could have a bearing on the cutoff level below which a risk is regarded as de minimis.

In developing an integrated philosophy regarding the introduction of de minimis risk concepts, attention needs to be directed to an appropriate interfacing of NRC's rules with the radiological protection rules of the EPA (40 CFR Part 190) as established in 10 CFR Part 20.106(g):

> (g) In addition to other requirements of this part, licensees engaged in uranium fuel cycle operations subject to the provisions of 40 CFR Part 190, "Environmental Radiation Protection Standard for Nuclear Power Operations," shall comply with that part.

Recently, NRC Commissioner James Asselstine raised a question regarding the relationship between the limits proposed in the Part 20 revision (SECY-85-147) and EPA's Uranium Fuel Cycle Standard (40 CFR Part 190). The staff responded that the two standards are quite different in their derivation and in the functions they serve.[14] Although both are radiation protections standards, one (Part 20) relates to *limitation of health risk*. The other (40 CFR Part 190) relates to the application of the ALARA principle involving an assessment of the cost and effectiveness of exposure *limiting technology* for specific types of fuel cycle facilities.

It has been the long-standing practice of the NRC and other organizations associated with radiation health protection to implement radiation protection by setting upper limits of exposure based on presidentially approved federal guidance and/or on ICRP and NCRP recommendations. But then actual exposures are kept as low as is reasonably achievable (or ALARA) below these limits. This approach was set forth in the Federal Radiation Protection Guidance issued in 1960:

> 1.18 Radiation Protection Guide (RPG) is the radiation dose which should not be exceeded without careful consideration of the reasons for doing so; every effort should be made to encourage the maintenance of radiation doses as far below this guide as practicable.

This latter terminology has evolved from "as low as practicable" (ALAP) to "as low as is reasonably achievable" (ALARA). The combination of exposure limits and lower ALARA exposures is today an inherent feature of all radiation protection programs.

The *limit* in the proposed Part 20 for exposure of the general public is 0.5 rem (500 millirem) per year. It limits the sum of the doses from all sources of radiation except background radiation and medical exposure to a patient. This limit derives from the existing Federal Radiation Guidance and is consistent with the recommendations of both the National Council on Radiation Protection and Measurements (NCRP) and the International Commission on Radiological (ICRP). This limit and corresponding health-based limits for occupational protection are considered upper bounds to acceptable doses and should be complied with regardless of cost. The *reference level* of 0.1 rem (100 millirem) per year in the proposed Part 20 defines a licensee action level and is consistent with recommendations in ICRP Publication No. 26 and with draft recommendations under consideration by the NCRP. It defines a level of radiation from a specific source. If this limit is exceeded, the licensee must conduct further investigations to determine that the combined exposure to individuals, including the radiation dose from the source in question, does not exceed 500 millirem per year. In addition, under the proposed Part 20, each licensee would be required to apply ALARA principals.

The health-risk-based levels in the proposed Part 20 reflect a judgment of the level of risk that would be acceptable in comparison with other risks. In contrast, EPA's Uranium Fuel Cycle Standard (40 CFR Part 190) and the reactor design objectives in Appendix I to 10 CFR Part 50 represent standards set on the basis of the quantified determination of what is "as low as is reasonably achievable," or ALARA. These levels are primarily technology-based. That is, they are based on analysis of technological capabilities and costs and the related reduction in offsite. population doses and potential health effects; they are based on what is both technically and economically achievable for specific types of fuel cycle facilities. The costs of achieving these standards are implicit in their development. As EPA stated in the environmental statement accompanying 40 CFR Part 190 (11, pp. 5–6):

> The environmental effects of planned releases of radioactive effluents from components of this cycle have been analyzed in detail by the EPA in a series of technical reports covering fuel supply facilities, light water reactors, and fuel reprocessing. . . . These technical analyses provide assessments of the potential health effects associated with each of the various types of planned releases of radioactivity from each of the various operations of the fuel cycle and the effectiveness and cost of the controls available to reduce releases of these effluents. In addition to these analyses, there is considerable other information on planned releases from these types of facilities available. This includes the generic findings of the NRC concerning the practicability of effluent controls for light-water-cooled reactors, extensive findings of the utilities, the NRC, and the AEC, reflected by environmental statements for a variety of individual facilities, and finally, the results of a number of detailed environmental surveys conducted by the EPA of typical operating facilities.

Appendix I to 10 CFR Part 50 contains design objectives defining ALARA effluent releases and, although it contains action levels, it does not contain dose limits in the usual sense. EPA's 40 CFR Part 190 does define dose limits for normal operation of uranium fuel cycle facilities, but these are not rigid, absolute limits, since the standard allows NRC to grant a variance to exceed these limits in cases where the need for continued operation has been demonstrated, and corrective action is being taken to reduce the exposures. The Appendix I design objectives (which preceded the issuance of the EPA's 40 CFR 190) are expressed separately for each reactor on a site and for individual exposure pathways (noble gases, radioiodines and particulates, and liquid effluents). The EPA standards in 40 CFR Part 190 apply to the entire site and to the total of all exposure pathways. When these differences are taken into account, the two standards are relatively compatible. This is not surprising since the EPA drew upon the same AEC studies and the operating experience used by NRC to develop Appendix I. It should be noted that the EPA 40 CFR 190 standard is incorporated into the present and proposed Part 20 by reference.

Control of radioactive releases and waste-processing technology for reactors and other fuel cycle facilities are very cost-effective and this results in very low ALARA levels. This is why it is appropriate that the EPA 40 CFR Part 190 standards (referenced in Part 20) and the levels attained by compliance with Appendix I are significantly lower than the proposed Part 20 limits and reference levels. However, the EPA standards in 40 CFR Part 190 specify ALARA-type levels only for uranium fuel cycle facilities and cannot generally be used as an indication of what is ALARA for other classes of NRC-licensed facilities. The important distinction is that the limit for members of the public in the proposed Part 20 limits the sum of the doses from all sources except background and medical radiation. In contrast, the EPA standard is an ALARA-based standard for uranium fuel cycle facilities and light water reactors. Establishing an upper boundary between acceptable and unacceptable risk is, of course, just the opposite of using ALARA procedures for determining a lower bound of de minimis risk below which cost-benefit analysis and further risk reduction is not required. The de minimis risk concept discussed above in the proposed revision to Part 20 does not serve this purpose but rather the more limited objective of establishing a cutoff value of 1 mrem per year to individuals in calculating collective dose from a given source.

On January 9, 1986, a modified version of the proposed NRC revision to 10 CFR Part 20 was published for public comment in the *Federal Register* (51 FR 1092). The new revision dropped the concept of applying a de minimis risk standard to the *most-exposed individual,* but retained the concept of applying a de minimis risk cutoff standard of 0.001 rem (or 1 mrem) per year for *collective dose calculations* with the following explanation:

> A more limited application of the de minimis concept has been proposed. Following consideration of lower and higher numbers, a value of 0.001 rem (0.01 mSv) per year was selected for limiting the extent of evaluating collective doses to populations. Application of the de minimis level to collective dose estimates would, among other things, limit both the size of the population and the time over which collective dose would need to be considered in evaluating activities associated with the release of radioactive materials to the environment.
>
> The proposed application of the de minimis concept could have a substantial influence on the evaluations of conditions where very large numbers of people are subjected to very low dose rates. In essence, the proposed rule would suggest disregarding extremely low dose rates (0.001 rem per year) without regard to the number of people exposed at that level or less. Thus, this contribution to estimates of collective doses would be disregarded. Where collective doses to a population are evaluated, the acceptability of the associated potential risks can also be compared to the sum of potential risks experienced by the same population over the same time interval. Consequently, even though some de minimis applications could result in very small but finite doses to very large numbers of persons, the comparative collective risk to which these people are routinely subjected (for example, from natural background radiation) is also very substantial and proportional to the number of persons considered.

Regulatory Exemption Levels*

An example of NRC's approach in deciding on exemptions to its radiological protection regulations relates to 10 CFR 20.106. This section requires licensees to assure that radioactivity in effluents to unrestricted areas does not exceed certain maximum permissible concentrations above natural background in air and water. Exemptions may be specifically authorized under that section or under section 20.302, relating to waste disposal. As noted above, section 20.106 also provides that licensees engaged in uranium fuel cycle operations shall comply with the "Environmental Radiation Protection Standards for Nuclear Power Operations," established by the U.S. Environmental Protection Agency in 40 CFR 190. Those standards set exposure limits of 25 millirems to the total body, 75 millirems per year to the thyroid, and 25 millirems per year to all other organs as a result of planned discharges to the environment from uranium fuel cycle activities. These levels of radiation exposure involve doses above which the Nuclear Regulatory Commission and the Environment Protection Agency consider the risks unacceptable, or at least unreasonable under normal circumstances. Although licensees are not required to reduce exposures or releases below the ALARA level, the Nuclear Regulatory Commission does not consider those exposures or releases trifling. In fact, further reduction would be

---

* Much of the material in this section has been adapted from a paper by Guy H. Cunningham, III, executive legal director of the U.S. Nuclear Regulatory Commission.[15]

required if circumstances changed to make a lower level *reasonably achievable*. In the absence of an explicitly established de minimis risk standard, there is no level below which further reductions might not, in principle, be required.

Under the Atomic Energy Act (the Act) the NRC has the authority to exempt from its regulations certain quantities or concentrations, forms, or uses of radioactive material. Under the Act, such material is classified as "source material," "by-product material," or "special nuclear material." Source material is defined as uranium, thorium, and their ores in designated concentrations, and may include any other material that is considered essential to the production of special nuclear material. Special nuclear material means plutonium, uranium enriched in the isotope 233 or 235, and any other material that the Nuclear Regulatory Commission may designate, but does not include source material. By-product material means any radioactive material (other than special nuclear material) yielded in or made radioactive by exposure to the radiation incident to the process of producing or utilizing special nuclear material. By-product material also includes uranium or thorium mill tailings. Under section 62 of the Atomic Energy Act, the Nuclear Regulatory Commission has the authority to exempt from its regulations unimportant quantities of source material. It also has authority, under Sections 57d and 81 of the Act, to exempt quantities of special nuclear material and by-product material, respectively, that would not be inimical to the common defense and security and would not constitute an unreasonable risk to the public health and safety. This exemption authority is reflected in various sections of the regulations: 10 CFR Part 40 for source material, 10 CFR Parts 30-35 for by-product material, and 10 CFR Part 70 for special nuclear materials.

Specific exemptions may involve exposures that are within the ALARA range but would not necessarily be de minimis, depending on the de minimis level chosen. Such exemptions usually permit use or disposal of exempted products or quantities without regard to their radioactivity. They are not properly considered de minimis levels, however, because they are not expressly based on a finding that the risk is negligible. Rather, exemptions codified in the regulation are based primarily on a balancing of benefit and cost, of which risk is an important element. There are several recent examples in which NRC amended several sections of the 10 CFR Part 20 regulations. On March 11, 1981, the commission amended Section 20.306 of its regulations to permit disposal of liquid scintillation media and animal carcasses containing tracer levels of tritium or carbon-14 (0.05 microcuries or less per gram) without regard to their radioactivity. It also amended Section 20.303(d) to raise the annual limits for disposal of tritium and carbon-14 by release to the sanitary sewerage system. That section now provides that each licensee may release into the sanitary sewerage system not more than five Curies per year of tritium and one Curie per year of carbon-14. These exemptions do not represent a complete deregulation of the disposal of such amounts of radioactive wastes, however. Disposal of liquid scintillation media and animal carcasses under Section 20.306 must be in accordance with all applicable federal, state, and local regulations governing any other toxic or hazardous property of those materials. Tissue may not be disposed of in a manner that would permit its use as food for either humans or animals. In addition, licensees are not relieved of maintaining records showing the receipt, transfer, and disposal of radioactive materials, as required by Part 30 of the regulations.

It is not clear what levels of societal risk are involved following these new permissible procedures or standards. Indeed, such calculations under these circumstances would be most difficult since exposure of humans to these sources of hazards is rather tenuous and highly uncertain. But it would appear that the risks are quite small and, moreover, the exemptions require no further application of ALARA to reduce these risks any further.

Thus, notes Cunningham,[15] we have a regulatory scheme with upper limits above which the calculated health risk is generally unacceptable. Below these upper limits are various specific provisions and exemptions involving calculated risks that are considered acceptable based on a balancing of benefits and costs, and these need not be considered further. Regulatory requirements below the upper limits are based on the ALARA principle, and any risk involved is judged acceptable given not only the magnitude of the health risk presented but also various social and economic considerations. A de minimis level, if adopted, would provide a regulatory cutoff below which any health risk, if present, could be considered negligible. Thus, the de minimis level would establish a lower limit for the ALARA range of doses.

Facility Decommissioning

In December 1978 a report (NUREG-0436, Rev. 1) was issued by the NRC that described a plan for reevaluation of NRC policy on decommissioning of nuclear facilities.[16] In addition to commercial nuclear power and research reactors, other nuclear fuel cycle facilities included in the decommissioning plan are uranium mills, UF_6 conversion plants, fuel fabrication plants, and fuel reprocessing plants. In addition, a study was initiated of the more than 20,000 licensed facilities that utilize by-product and source materials (e.g., radioisotope applications), which represent a rather limited decommissioning problem because they handle small amounts of radioisotopes.

The nuclear field is reaching the degree of maturity that requires increased attention to the proper retirement or decommissioning of facilities. Thus, more nuclear plants and equipment will presently be entering the terminal period of their useful lives. Since most of these facilities have been involved in handling radioactive materials, emphasis in policy development is placed upon the safety of the decommissioning process, including handling and disposal of residues. As stated in Regulatory Guide 1.86, decommissioning alternatives presently acceptable to the staff include mothballing, entombment, prompt removal/dismantling, or a combination of any of the three.[17] Protective storage or mothballing involves the removal of all fuel and source material, the disposal of all liquid and solid waste, and placing the facility in a state of protective storage. Entombment requires similar treatment, and, additionally, the radioactive materials and components are encased (usually in concrete) and isolated until they decay to unrestricted levels. Removal and dismantling require that all radioactive structures, components, and systems be disposed of such that the site can be released for unrestricted use. In each licensing case the staff must be satisfied that a feasible decommissioning method exists and that the applicant possesses, or has the capability to provide, the necessary funds to complete the task.

At termination of nuclear reactor operation it may be advantageous to the licensee to request a possession-only license which permits possession of the facility and remaining radioactive materials but continues to impose requirements upon the licensee to ensure that proper surveillance and procedures are followed to maintain the nonoperating facility in a safe condition. The licensee remains responsible for the safe disposition of fuel, radioactive components, and other radioactive source material. Radiation monitoring, environmental surveillance, and appropriate plant security procedures are required for an indefinite time into the future to ensure that the public is not endangered.

Since residual quantities of source materials (and especially of surface contaminants of radionuclides) are often quite small in the decommissioning alternatives for nuclear facilities, the concept of establishing a de minimis standard of acceptable risk seems quite appropriate. Indeed, the earliest explicit attention to the de minimis risk concept that I have been able to uncover in NRC's development of regulatory policies is found in the following passage from the 1978 publication of NUREG-0436:

> 5.3 *Principal Issues in Decommissioning Policy.* The principal issues to be addressed in developing or reevaluating decommissioning policy are the acceptability of radioactive residue levels (i.e., "de minimis" criteria) and the financial assurance. Much of the information needed for our policy development can come from the work of contractors, but in these principal areas direct and particular attention by the NRC staff is very important.[16]

However, this report did not propose any quantitative level of de minimis risk, noting only that the lack of any authoritative definition of a "de minimis" dose (i.e., low-dose equivalent in risk to other activities that are generally accepted without special concern) also meant that a proposed acceptable level of residual radioactivity in soil—and presumably also for surface contamination on facility hardware—was certain to be in jeopardy. An appreciation of the difficulties in establishing in 1974 the numerical acceptable surface contamination levels is shown in Table 1 (Reg. Guide 1.86, p. 5). These standards are still utilized today for reviewing a facility dismantling plan and termination of a facility license. Thus, if residual radiation levels do not exceed the values in Table 1, the commission may terminate the license. If these levels are exceeded, the licensee retains the possession-only license under which the dismantling activities have been conducted or, as an alternative, may make application to the state (if an agreement state) for a by-product materials license.

Among the many difficulties in determining the quantitative health effects implications of the acceptable surface contamination levels of Table 1 (which are readily metered) are these:

- There are a legion of different mixes of alpha, beta, gamma rays, etc., and their disparate energy levels that could produce the same radiation level as measured in the number of disintegrations per minute.* The depth of penetration in human tissues and organs is a function of the type of radiation and their associated energy levels.
- There are a variety of radiological pathways by which human exposure could result in harm (e.g., dermal contact or inhalation with fugitive dust or particulates, inhalation of gaseous radionuclides, ingestion of contaminated food and drinking water, and others).
- Variable population densities exist at different distances from the surface contaminants and their daughter radionuclides that could become airborne or waterborne, and these also exists the unpredictability of future levels of human activities.

---

*One Curie is equal to 3.7×10^{10} disintegrations per second.

Table 1. Acceptable Surface Contamination Levels

Nuclide[a]	Average[b,c]	Maximum[b,d]	Removable[b,e]
U-nat, U-235, U-238, and associated decay products	5,000 dpm α/100 cm^2	15,000 dpm α/100 cm^2	1,000 dpm α/100 cm^2
Transuranics, Ra-226, Ra-228, Th-230, Th-228, Pa-231, Ac-227, I-125, I-129	100 dpm/100 cm^2	300 dpm/100 cm^2	20 dpm/100 cm^2
Th-nat, Th-232, Sr-90, Ra-223, Ra-224, U-232, I-126, I-131, I-133	1,000 dpm/100 cm^2	3,000 dpm/100 cm^2	200 dpm/100 cm^2
Beta-gamma emitters (nuclides with decay modes other than alpha emission or spontaneous fission) except Sr-90 and others noted above	5,000 dpm β-γ/100 cm^2	15,000 dpm β-γ/100 cm^2	1,000 dpm β-γ/100 cm^2

Source: USAEC Regulatory Guide 1.86.[17]

[a] Where surface contamination by both alpha- and beta-gamma-emitting nuclides exists, the limits established for alpha- and beta-gamma-emitting nuclides should apply independently.

[b] As used in this table, dpm (disintegrations per minute) means the rate of emission by radioactive material as determined by correcting the counts per minute observed by an appropriate detector for background, efficiency, and geometric factors associated with the instrumentation.

[c] Measurements of average contaminant should not be averaged over more than 1 square meter. For objects of less surface area, the average should be derived for each such object.

[d] The maximum contamination level applies to an area of not more than 100 cm^2.

[e] The amount of removable radioactive material per 100 cm^2 of surface area should be determined by wiping that area with dry filter or soft absorbent paper, applying moderate presssure, and assessing the amount of radioactive material on the wipe with an appropriate instrument of known efficiency. When removable contamination on objects of less surface area is determined, the pertinent levels should be reduced proportionally and the entire surface should be wiped.

- There are scientific uncertainties in dose–response modeling or the use of epidemiological data to infer cancer fatalities from given dose levels for different types of radiation, especially for low doses.[18–20]

   Given the nature of these variables, one would be hard-put to impute the health effects corresponding to the acceptable surface contamination levels of Table 1. For example, radioactive fallout from weapons testing led to a peak value of about 5 mrem per year of exposure in the early 1960s but has declined substantially since then. As noted above in the discussion of 10 CFR Part 20 standards, 5 mrem per year of whole body dose is roughly equivalent to a cancer fatality rate of 6.7×10^{-7} deaths per year of exposure. According to common-use interpretations of insignificant risk to the individual, 5 mrem per year could be regarded as de minimis were it not for the unresolved problem of scientific uncertainty. Whether the extent of contaminated surfaces from facility decommissioning would reach this level of radioactive exposure to large regions of the country at the acceptable levels of Table 1 is an unresolved question.

   Also, the setting of acceptable residual levels of contamination requires a decision on what credit can be taken, and for how long, for isolating the contamination by direct custody (guards, fences, etc.), or by confining media (paint, concrete, soil cover, etc.). It is an underlying assumption that NRC will actively seek to prevent a proliferation of sites where the remains of nuclear facility operations stand guarded or entombed. At the same time, the NRC recognizes that some sites, such as major power plant sites, are uniquely suited to this purpose for the foreseeable future, and it may not be reasonable for our society to expend substantial resources and increase doses to the workers to decommission an old nuclear facility site to a pristine condition, if a new nuclear facility is to be built there. The above examples focus on the level of individual risk associated with only one pathway of radiological exposure, ignoring other pathways from recycling contaminated materials, etc. Thus, the question of the aggregative individual or societal risk from multiple pathways has not yet been explored in this context.

In conjunction with the development of a plan for reevaluation of NRC policy in decommissioning nuclear facilities, a number of state workshops were held in 1978, plus others since then. Regarding the determination of a level of individual risk in the common-use range of de minimis risk, these workshops were asked to address the following specific question (reference 16, p. F-11): "Is a maximum dose rate of 1 mrem per year to any individual after cleanup an acceptable basis for site release? What other basis would you recommend?" The prevailing view was that an "acceptable and reasonable standard must be set. The rationale for choosing the standard must be well documented."

Specific points discussed were as follows:

- This standard is unrealistic (4 groups).
- A range is more appropriate: minimum 1 mrem–maximum 5 mrems (2 groups).
- Different dose rates should be set for different types of facilities; it should be set on a site-by-site basis.
- If it is achievable on a cost-effective basis for site release, then it is acceptable.
- It should be related to risk assessment.
- This standard should be set by EPA.
- A reasonable maximum dose rate that is nationally accepted should be established with the states' right to set stricter exposure levels on a case-by-case basis.
- The group didn't have enough information—but listed concerns that federal experts should study:
  - Concerned with problem of defining background including any changes over the life of the facility.
  - Concerned with enforceability because so close to background.
  - Concerned with the size and type of the population that would be affected.
  - Concerned that the standard should reflect the duration of exposure.
  - Concerned with the relation of the maximum dose rate to a health effect.
- In determining a more reasonable number for use, options such as a percentage of background, a sliding percentage (10 CFR 50, App. I), or use of levels presently set for exposure in operating reactors should be investigated.

Following the state-level workshops, a public meeting was held in Washington, D.C., in accordance with the announcement in the *Federal Register* notice of August 25, 1978 (43 FR 38025). Attendance included representatives of the nuclear industry, public interest groups, government agencies and the states. As reported in NUREG-0436 (Rev. 1), one of the conclusions was this:

In regard to residual radioactivity levels (i.e., "de minimis" values*) it was commented that the dose values of 1 to 25 mrem, as used as an illustrative example in the NUREG reports presented, were too low and impractical because of difficulty in measurements at these two levels. (reference 16, p. 35)

As previously noted, the acceptance surface contamination levels stated in Table 1 currently remain the applicable NRC numerical standards for facility decommissioning. However, the NRC is reevaluating its decommissioning policy, a process that began with the preparation in 1981 of the Draft Generic Environmental Impact Statement on Decommissioning of Nuclear Facilities (NUREG-0586). The Draft GEIS dealt with a wide scope of nuclear fuel cycle facilities but excluded from consideration for regulation change the decommissioning of shallow-land low-level waste burial grounds, deep geologic high-level waste burial, and uranium mill and mill tailings, which are being considered in separate rule-making activities, and the decommissioning of uranium mines that are not under NRC jurisdiction. Recommendations were made as to the regulatory decommissioning particulars, including such aspects as appropriate initial planning requirements prior to commissioning, final planning requirements level for unrestricted access, and assurance of funding for decommissioning.

According to the Draft GEIS, an important and technically difficult issue is the problem of determining acceptable residual radioactivity levels required for release of property for unrestricted use. It is the responsibility of the Environmental Protection Agency (EPA) to establish such a standard; this is currently being attempted by the EPA. Discussions have been held with the EPA relative to providing guidance for establishing limits that are consistent with eventual EPA requirements. Owing to the variety of facility types and radionuclides involved, it is not feasible to set a single dose limit that would be valid under all conditions for all facilities. It is necessary to assess the radiological impact in terms of the radionuclides and pathways involved and the costs and benefits that result. In the light of discussions with the EPA and considerations that the level of residual radioactivity selected must be safe and consistent with existing guidance and be measurable and cost effective, the following results were determined:

1. A residual radioactivity level for permitting release of a nuclear facility for unrestricted use should be ALARA. Guidance in establishing such a limiting level is best expressed in terms of a value that bounds the dose for the majority of facilities discussed in this report. This value is determined to be 10 mrem per year whole-body dose equivalent, but it could be lower for specific facilities. The 10-mrem-per-year limit is chosen recognizing that it may be impractical and unnecessary in some cases to meet a 5-mrem-per-year limit considered in previous discussions with EPA. This is because of cost-benefit considerations and problems in detectability, sampling, and/or exposure patterns. Discussion with EPA indicated that the 10-mrem-per-year limiting value would not be considered unreasonable. In all cases, a dose limit above 1 mrem per year would require justification. For a few situations, it is expected that residual limits will be outside the bounds of the 1- to 10-mrem per year range. For these special situations,

---

* *De minimis* levels are those levels of radioactive contamination that are so low that the site or object so contaminated may be released without further concern or restriction—taken from the Latin: *de minimis non curat lex* ("the law does not deal with trifles").

case-by-case analysis in terms of cost and benefit effectiveness will be required to establish appropriate limiting levels.

2. For implementation of a residual radioactivity level, the dose value selected must be converted to a contaminated material concentration or activity for instrument measurability. Such conversion is done through the use of modeling and depends on what radionuclides are present and how they result in individual radioactivity exposure. Realistic exposure conditions should be used in such modeling, recognizing, for example, the dwelling occupancy is less than full time, that self-shielding is an important exposure-reducing factor, and that weathering reduces resuspension of the contaminated materials.

On February 11, 1985, the NRC published for public comment a proposed rule, "Decommissioning Criteria for Nuclear Facilities" (50 FR 5600). Decommissioning as defined in this proposed rule means to remove nuclear facilities safely from service and reduce residual radioactivity to a level that permits release of the property for unrestricted use and termination of license. For the purpose of this proposed rule, the term *nuclear facilities* is used to refer to the site, buildings and contents, and equipment associated with any NRC licensed activity. To release property for unrestricted use, a permissible level of residual radioactivity must be established. These levels are not proposed in this rule but are being developed in a separate rulemaking for facility decommissioning criteria. However, the proposed planning requirements are considered appropriate means of assuring that the decommissioning will be carried out in accordance with Part 20 and specifically that doses will be kept as low as reasonably achievable.

Uranium Mines and Mills

The radiation doses to the public at distances of several miles or more from a uranium mill and mill tailings site are sufficiently low as to invite consideration of using de minimis risk concepts in their regulation, or at least the 1-mrem-per-year cutoff concept in aggregate population dose calculations as currently proposed in the revisions to 10 CFR Part 20 (see above) for other types of nuclear fuel cycle facilities. It is fair to say there is a significant rationale for both sides of the argument. Fleishman of the UK's National Radiological Protection Board provides a logic for both the individual and societally aggregate risk perspectives:

> [T]here has been a protracted debate as to whether or not it is reasonable to sum very small individual doses, additional to those accumulated from natural background radiation, in the calculation and use of collective dose estimates. This issue has assumed considerable importance in the field of radioactive waste management, as much of the collective dose arising from the dispersion of effluents or low-level solid waste disposal is delivered at very low individual dose rates. A widespread feeling has emerged within the radiological protection community that this procedure may over-emphasize the significance of low-level exposure and, in particular, promote undue public anxieties over those waste management practices that result in essentially trivial levels of risk. As an alternative, it has been suggested that a level of individual dose or risk should be established below which there would be no further need for regulatory concern. Such dose levels or corresponding levels of risk are often termed *de minimis*. . . . However, there are two features which cause difficulty with this approach. One is the likelihood that a particular individual will be exposed to many sources, each of which may have been dismissed as of no

concern, and the second is that one source, which may have been dismissed on an individual basis, will in fact expose many individuals.[21]

Although the regulatory analysis leading to an NRC decision on the treatment of the radiological risks from mill tailings following decommissioning did not explicitly discuss nor establish a de minimis criterion, the discussion to follow reveals the sort of issues and data analysis provided by the staff and licensing boards that would be most relevant to any future development of this kind.

Management of mill tailings after milling operations cease, involving about 600,000 tons per year of material with low radionuclide concentrations, is the major consideration in decommissioning of these facilities. The major problem encountered in past milling operations is the management of tailings generated by the milling process. Although the concentration of radioactivity in the tailings is relatively low, control measures are necessary because of the large quantities involved and because of the long half-life of the parent radionuclides that are present. The management of mill tailings has received increasing attention and interest in recent years from involved federal and state agencies and from environmental conservation groups. This interest has resulted from studies carried out during the last decade which have indicated that uranium mill tailings, if not properly managed and controlled, could present a potential public health hazard. The most vivid example, of course, is the situation that occurred in Grand Junction, Colorado. The remedial actions determined necessary to correct the misuse of tailings in the construction of homes, schools, and other public structures are continuing at substantial cost to the federal government and the state of Colorado.

The issues and alternatives for providing acceptable radiological protection to the public were examined in detail by the NRC Office of Nuclear Material Safety and Safeguards. The Final Generic Environmental Impact Statement (GEIS) on Uranium Milling, published in September 1980, provides a benefit-cost analysis of both radiological and nonradiological impacts.[22] The key objective of the GEIS are (1) to assess the nature and extent of the environmental impacts of conventional uranium milling in the United States from local, regional, and national perspectives on both short- and long-term bases to determine what regulatory actions are needed; (2) more specifically, to provide information on which to determine what regulatory requirements for management and disposal of mill tailings and mill decommissioning should be; and (3) to support any regulation changes that may be determined to be necessary.

Specific regulatory changes found to be needed as a result of the analysis performed were issued simultaneously with the GEIS. In addition, the regulations incorporate requirements of the Uranium Mill Tailings Radiation Control Act of 1978 as amended. Requirements in the regulations affecting emissions during operation will assure that exposures to individuals are within existing public health standards. Furthermore, requirements of regulations will have the effect of assuring that mill operations are performed in a manner that reduces population exposures and risks to the maximum extent reasonably achievable.

With respect to overall health impacts, the critical mill-released radionuclides and their primary sources are, in descending order of importance, radon-222 from the tailings pile, radium-226 and lead-210 from the tailings pile, and uranium-238 and uranium-234 from yellowcake operations. Health impacts from radon-222 result from inhalation of ingrown daughters and ingestion of the ground-deposited, long-lived daughter lead-210. Because radon-222 is released in gaseous form, it is transported long distances exposing large populations, albeit at extremely small levels above background. The impacts of radium-226 and lead-210, released in particulate form from the tailings pile, result primarily through ingestion pathways (dispersed radium-226 also constitutes a secondary source of radon-222 release). Emissions from impounded tailings materials have an enhanced importance due to their persistence beyond the operational lifetime of the mill itself. Yellowcake emissions result in significant localized impacts, primarily via inhalation, but essentially terminate when the mill shuts down.

The NRC proposed regulations require changes in current milling practices and measures to reduce long-term societal risk from abandoned mill tailings that will conform to the EPA radiological protection standards in 40 CFR Part 190 and 40 CFR Part 192. The EPA standards require that exposures of whole body or any organ to any individual in the general public not exceed 25 mrem per year. Because the limits apply only to exposure to nuclides other than radon and its daughters, the first exposures discussed (referred to as 40 CFR 190 doses) do not include contributions from these nuclides, which are regulated under 40 CFR Part 192. Total exposures to the individuals and populations also discussed include doses resulting from the radon component, and health risks associated with these total exposures.

Under the former regulations, the 40 CFR Part 190 limits were not met at permanent residence locations near uranium mills. Doses received by the nearby individual substantially exceeded 25 mrem per year; bone and lung doses are 45 and 30 mrem, respectively. Analysis indicates that the limit could not be met within about 3 km downwind from the mill. The effect on the nearby resident of a potential worst case concentration of milling activity, where a cluster of 12 mills is postulated in year 2000, would be to increase 40 CFR 190 doses to bone and lung by about 15 to 20%. Although not a large fractional increase, this shows that the contribution from surrounding mills could be important in situations where meeting 40 CFR 190 was otherwise a borderline case.

According to the GEIS, total exposure estimates (which include radon and daughters) indicate that radon is the greatest single contributor to risk. When total exposures are considered, the chances that the nearby individual would prematurely die from cancer as a result of living near the model mill for 20 years (a period assumed to include the full operation and decommissioning cycle of the mill) would be about 380 in a million. Thus, at a rate of 19 cancer fatalities in a million per year of exposure, this is substantially above the 10^{-5} to 10^{-6} annual fatality risk level commonly suggested as a standard for individual de minimis risk. Because of the considerable uncertainties that exist in the health risk estimates used (risks could be one-half to two times those estimated), comparison with risks posed by background radiation provides valuable perspective. The estimated risks to the nearby individual would be an increase of about 25% above risks from background radiation exposures. Exposures and risks to an average individual in the region over a similar time span would be a small fraction (less than 1%) of those for the nearby individual.

The effect of concentrated milling activity would be to increase risks to the nearby individual by about 50%. The milling cluster would have a more dramatic effect on risks for the average individual, raising them by a factor of about 10, from 3 to about 30 chances in 1 million of premature cancer death. This risk would be about 2% of that faced owing to natural radiation exposures. The risk to an average individual living in a region of maximum mining and milling activity in the year 2000 is very roughly estimated to be double those described above for milling alone. This estimate is based on recent radon measurements around open-pit and underground mines which indicate that releases from active mining will be roughly equivalent to those that would occur from tailings under the base case.

A useful perspective on radiological impacts on workers and nearby individuals of the general public from radioactive airborne emissions for the GEIS base-case model mill is shown in Table 2.[22] Radiological impacts will depend on the number of mills in a region and on the mill size and other variable characteristics of mill design and operation. Accordingly, the GEIS developed a generic description of a "model mill" based on features typical of uranium mills in operation in the early 1970s. The characteristics, operating procedures, processes, and effluents of the model mill were derived from data for existing mills that are described in technical literature and various environmental reports and statements. The model mill concept serves two basic functions: (1) It provides a means of assessing the environmental impact of the model region and the model site, and (2)

Table 2. Radiological Impacts from Radioactive Airborne Emissions for the Base Case Model Mill

Receptor	Dose commitment[a] (mrem)			Risk from mill as percentage of risk due to background (%)[b,c]
	Whole body	Bone	Lung	
Nearby individual[d]				
Annual 40 CFR 190 doses				
(excluding radon)				
1 mill	3	45	30	—
Mill cluster	4	51	36	—
Total dose				
(including radon)				
1 mill	9.7	51	220	25
Mill cluster	13	61	340	38
Average individual[e]				
1 mill	0.061	0.50	1.6	0.19
Mill cluster	0.66	5.8	16	1.9
Average worker[f]				
Annual	450	2000	7100	800
Career[g]	2.1×10^4	9.3×10^4	3.3×10^5	800
Background	143	250	704	—

Source: NUREG-0706 (22).

[a] All doses shown are total annual 15th-year dose commitments except where noted as being those covered by 40 CFR 190 limits.

[b] The range in risks due to uncertainties in health effects models extends from about one-half to two times the central value (App. G-7). This range does not include uncertainties in other areas (e.g., source term estimates and dose assessment models).

[c] Risk comparisons are presented for exposure received during entire mill life; that is, 15 years of exposure during operation of the mill, and 5 years of postoperation exposure while tailings are drying out are considered. This value is greater than that from annual exposures presented because tailings dust releases increase in the period when tailings are drying.

[d] The "nearby individual" occupies a permanent residence at a reference location about 2 km downwind of the tailings pile.

[e] The "average individual" exposure is determined by dividing total population exposure in the model region by its population total.

[f] The "average worker" exposure is determined by averging exposures expected at the various locations in the typical mill.

[g] The career dose is based on a person who has worked 47 years in the milling industry (that is, from ages 18 to 65).

it serves as a base case for evaluating the environmental impacts of alternative methods of effluent control and tailings management. The model mill features a relatively low level of environmental control, which in some respects represents a lower level of control than that currently used at U.S. mills.

Depending on the chemical characteristics of the ore, conventional uranium mills employ either the acid-leach process coupled with solvent extraction or the alkaline-leach process coupled with caustic precipitation for the concentration and purification of leached uranium. These processes are most common in the industry at present, and this situation is expected to continue for the period of interest. As of 1976, mills employing the acid-leach process represented 82% of the total U_3O_8 production capacity of the conventional milling industry; mills with alkaline-leach circuits accounted for the remaining 18%. In view of the preponderance of acid-leach mills, the model mill employed the acid-leach process. However, the major impacts from the alkaline-leach process are not expected to be significantly different from those of the acid-leach process. The basic assumptions in defining the model mill characteristics are shown in Table 3. Other features of this generic approach in defining the model mill and related source term and other assumptions important to radiological impact assessment are found in the GEIS.[22]

In comparing the dose commitments leading to health effects impacts (i.e., latent

Table 3. Summary of Principal Operating Characteristics of the Model Mill

Parameter	Value
Ore process rate	1800 MT/d
Average ore grade (% U_3O_8) 1980–2000	0.10%
Ore activity U-238 and each daughter in secular equilibrium (0.10% U_3O_8)	280 pCi/g
Ore transport	Haulage from mine to mill by truck (23 MT average load per truck)
Ore hauling distance	15 to 80 km (50-km average)
Ore pad area normally in use	0.5 ha
Ore storage time (10-day supply on hand)	~ 12 d
Operating days per year	310
Manpower requirements	~ 160 employees and other onsite
Uranium recovery (extraction efficiency)	93%
Product purity	90% U_3O_8
Average annual production (0.10% U_3O_8)	520 MT U_3O_8 or 580 MT yellowcake
Yellowcake transport	Shipment in 55-gallon drums by truck; each drum containing a maximum of 430 kg of yellowcake; 40 drums carried per truck
Dry solid waste generated (tailings)	1800 MT/d
Tailings density (slurry)	1.6 g/cm^3
Gross water flow to tailings pond	1800 MT/d
Tailings pond water recycled	30%
Net water consumption for tailings slurry	1260 MT/d
Area of milling facility (excluding tailings pile)	50 ha
Area of tailings impoundment	100 ha
Extra unused land	150 ha
Total area owned by milling operation	300 ha

Source: NUREG-0706 (22).

cancer fatality risks) in Table 2 for total dose including radon with those excluding radon, it is seen that a major difference exists for lung dose and to a lesser extent for whole-body dose. The estimated lung dose, including radon for a single mill, is 220 mrem, or 7.3 times the level excluding radon. The whole-body dose including radon (at 9.7 mrem) is about 3.2 times the dose excluding radon. In contrast, the bone dose commitment including radon is only 12% higher than that excluding radon.

The integrated risk assessment for an off-site individual in Table 2 for cancer fatality rates using model mill assumptions is given as 25% of background risk for a single mill and 38% for a mill cluster. There is a significant basis for the view that a one-third or so increase in radiation dose relative to natural background radiation might eventually be regarded as a de minimis level of individual risk.[23] For example, the average available background radiation dose in Colorado from cosmic and terrestrial radiation is 124 mrem per year. This is 2.4 times the 52 mrem per year estimated for the lowest state (Florida). Richard Wilson points out that a resident in a state with a low level of cosmic background radiation increases his or her risk of premature fatality by 1 part in 1 million during a two-month visit to Denver, the same as for the increased level of dosage from cosmic radiation in a jet flying for 6000 miles at 35,000 ft. altitude.[2] Moreover, a comparison of state data for annual cancer fatality rates with estimates for natural background suggests, if anything, an inverse correlation at these low rates of radiation doses. For example, the six highest states in terms of total cosmic and terrestrial radiation (Colorado, Montana, Wyoming, Utah, Idaho, and New Mexico) with estimates ranging from 88 to 124 mrem per year have cancer death rates of 91 to 162 per 100,000 population that are appreciably *below* the U.S. annual rate of 187. It is interesting to note that four of these states (New Mexico, Wyoming, Utah, and Colorado) have many active uranium mines and mills and are estimated collectively to possess 70% of the probable uranium resources of the United States (reference 22, pp. 3–15).

This negative correlation is not to suggest that low doses of radiation might be beneficial to one's health, but rather it is most likely due to the masking effects of far more important causal contributions to cancer fatality rates that would explain variations between states. These include demographic and socioeconomic factors, dietary habits, smoking and other life-style differences, and environmental chemical and radiation carcinogens of artifactual or natural origin.[23]

Analysis of cancer fatality data from the large sample of A-bomb survivors of Hiroshima and Nagasaki, as well as smaller subpopulations of industrial workers and of medical patients exposed to substantial doses of radiation, suggests that a dose of 10 rem is still sufficiently small enough that detection of an increase above the normal background incidence of cancer cannot generally be demonstrated with statistical precision even in a large exposed population.[24,25] Thus, the jury is still out on whether the levels of radiation doses estimated for the model mill cases in Table 2 might ultimately be accepted as de minimis risk according to common-use concepts of individual exposure.

The potentially most useful, and possibly controversial, application of a de minimis risk standard would be as a cutoff concept for low levels of risk posed by mill tailings for future generations as well as for long distances from mill tailing sites. In principle, the relevant ethical issues are not wholly dissimilar from those associated with multi-generational effects of long-term storage of high-level wastes from spent fuel of nuclear power plants that have been widely discussed.[26,27] However, since the ratio of the tonnage of mill tailings produced is about 10,000 times as large as the spent fuel in generating electricity, the spent fuel can affordably be buried at depths in the earth that would merit much less concern than mill tailings for cumulative effects on future generations.

The following estimates in the GEIS provide helpful perspective on the risk issues related to mill tailings:

- The most significant impact from mill operations under the base case would occur from persistent radon releases from the uncovered tailings. About 6000 premature deaths are calculated over the period 1979 to 3000 in the United States, Canada, and Mexico from tailings which would be generated by the full operation of mills in the U.S. through the year 2000.
- The cumulative potential impacts constitute a 1.2×10^{-5} fraction of the overall U.S. incidence of cancer. This level of average (individual) risk is almost two orders of magnitude (i.e., 80 times) less than the average level of latent cancer fatality risk to the population in the vicinity of a nuclear power plant from a severe accident, as stated in NRC's proposed safety goal—namely, a 10^{-3} fraction of background latent cancer fatalities (see below).
- Furthermore, the effects of releases from milling can be compared with those occurring from natural and technologically enhanced sources of radon. Specifically, exposures from milling radon releases would be about 0.3, 0.2, 3, and 10% of exposures occurring from releases from natural soils, building interiors, evapotranspiration and tilling of soil, respectively.
- The continuing annual rate of premature deaths from this volume of tailings is estimated to be about six per year. This annual rate could be used to develop estimates of health effects beyond 1000 years if this were desired; this would require making very uncertain assumptions on long-term factors such as climate, population growth, and the like (reference 22, p. 5).

As noted above, the base case for these estimates pertain to milling operations in the 1970s and do not reflect current protective measures to reduce these risks under the new regulatory requirements. In developing these requirements, the GEIS used risk-cost-benefit balancing that sought to address the following issues (reference 22, Vol. III, pp. U-6, U-7): (a) Should available societal resources be committed to protection from tailings-emanated radon when greater benefits might derive from other applications? (b) Should health risks occurring far in the future be valued as though they were occurring now? And if not, how should future health risks be valued? (c) Is it really worth reducing risks that, on an individual basis, are at least extremely small and may even be zero? (d) Should present generations be permitted to bequeath a legacy of continuing undeserved radiation exposure?

The dollar values and ranges of parameters used in this risk-cost-benefit analysis of alternatives for increased radiological protection are shown in Tables 4 and 5 (reference 22, Vol. III, p. U-6). After grappling with the above ethical issues in the sobering context and complexities of risk-cost-benefit analysis, the GEIS states (reference 22, Vol. III, p. U-7):

Table 4. Values and Ranges of Parameters Used in Radon Flux Optimization

Parameter	Definition and units	Central value	Range
A_1	Health effects/yr per pCi/m^2-s of Rn released	0.021	0.0021[a] to 0.21
A_3	Millions of $ per meter of earth cover applied	106	76 to 320
A_4	Radon flux attenuation coefficient, per meter	1.3	0.8 to 3.0
A_5	Millions of $ per health effect	0.4[b]	0.02 to 10
T	Effective number of years of annual health effects counted	—	100 to 100,000[c]

Source: NUREG-0706 (22).
[a] An ultimate lower limit of zero is possible.
[b] Central values given here for other parameters result directly from staff analysis and are correctly interpreted as staff "best estimates." The central value shown for the dollar worth of a health effect merely represents a middle value in a wide range of values appearing in the open literature; it is not otherwise adopted or endorsed by the staff.
[c] An ultimate upper limit of 4.5×10^8 has been determined.

These questions are, quite simply, beyond technical resolution. Although pertinent and important, they so involve emotional, political, and even religious considerations that the application of purely scientific knowledge and analysis is of no avail. Perspectives on how these questions should be answered varied widely among commentors on the draft GEIS. Many stated that the tailings isolation problem should be properly viewed as a short-term one when weighing the amount of radon control that should be required; for example, 100 years was urged as the period over which health effects should be integrated. Other commentors urged that the full period of toxicity should be considered and, based on this and the large number of potential health effects, virtually no radon releases from the tailings should be allowed.

Notwithstanding the enormous subjectivity involved with these questions, some might be willing to make judgments about them and select values for the parameters described in Section 2 and 3 that, for example, lie somewhere between the extremes. The optimization methodology would still break down for the case of the uranium mill tailings disposal problem, however. This is largely due to the impossibility of correlating containment performance uniquely with all applicable costs (that is, relating x to parameter A_2 as discussed in Section 3.2).

Table 5. Results Obtained in Radon Flux Optimization

Value of parameter T (yr)	Calculated optimum radon flux limits (pCi/m^2-s)	
	Central value	Range[a]
450,000,000	0.000022	0.00000003 to 0.021
100,000	0.097	0.00012 to 95
10,000	0.97	0.0012 to unlimited
1,000	9.7	0.012 to unlimited
100	97.0	0.12 to unlimited

Source: NUREG-0706 (22).
[a] Values shown are those obtained by minimizing or maximizing the optimum flux with respect to all parameters except T, by using parameter ranges listed in Table 4. The ultimate minimum value of parameter A_1 may be zero, in which case no radon flux control would be optimum, regardless of the values of other parameters. "T" represents different assumptions of cutoff periods for estimating health effects for future generations.

Despite these difficulties, the GEIS did reach certain regulatory conclusions and developed new requirements for measures costing tens of millions of dollars to increase radiological protection beyond former practices. The details of the alternative measures considered in the analysis and the implications of the new requirements for different types of mining and milling operations and environmental situations are beyond the scope of this chapter.* However, the resolution of these regulatory requirements left open the question of a future commission position on de minimis risk and whether such a concept should be established at the level of individual de minimis risk or the aggregate level of societal risk.[13,28]

According to Cunningham,[15] recognition of the importance of the de minimis risk concept emerged in a series of licensing decisions concerning the release of radon gas attributable to the mining and milling of uranium needed to fuel certain nuclear power reactors.** The issue in those cases arose under the National Environmental Policy Act rather than the Atomic Energy Act. Because the decisions concerned small release of radioactive radon gas, they are closely related to the issue of a de minimis concept in radiation protection.

In a proceeding to determine whether permission to construct a nuclear facility should be granted, the Nuclear Regulatory Commission's licensing board considered the significance of radon gas releases in the context of an environmental cost-benefit analysis. The licensing board found that the calculated health effect that might result—half a death per year in a population of 300 million—was a minimal impact. Properly stabilized mill tailings piles and reclaimed uranium mines would make the impact 100 times lower. The board concluded that the best means of characterizing the significance of radon releases attributable to operation of the facility was to compare them with those associated with natural background radiation and its fluctuations. The incremental releases attributable to the facility were so small as to be completely undetectable. Thus, the board concluded that their impact could not be significant, 8 NRC 87 (1978).

In several other licensing cases involving the same issue, the Nuclear Regulatory Commission's appeal boards referred to the licensing board's decision as employing a "de minimis approach," 13 NRC 487 (1981). In a more recent decision concerning the health effects of radon gas releases, the appeal boards made no reference to the de minimis rationale. They retained the comparison to natural background radiation, however, and concluded that "the *incremental* health risk to the population stemming from the fuel cycle emissions (if indeed there is any) is vanishingly small," 16 NRC 1528 (November 19, 1982). The appeal boards found that the radon releases attributable to a single 1000-megawatt (electric) nuclear reactor would cause an increase in dose to the bronchial epithelium of from 0.0005 to 0.005 millirem per year. This is far below the 480 millirem per year delivered to the bronchial epithelium by typical outdoor radon levels from background sources.

Nevertheless, the Nuclear Regulatory Commission has squarely rejected the argument

---

* On May 20, 1985, the NRC staff sent to the commission for its review and approval the proposed "Final Amendments to Uranium Mill Tailings Rule To Implement EPA Stability and Radon Release Standards" (SECY-85-178).

** The remainder of this discussion of de minimis risk in the NRC's regulatory treatment of mill tailings issues and the following section on nuclear waste management is adapted from Cunningham's paper.[15]

that the risks posed by uranium milling are de minimis. In denying a stay of its licensing requirements for uranium milling, sought by several mill operators and the state of New Mexico, the Commission explained:

> Radon is the primary source of long-term public exposure to radiation resulting from uranium milling. If adequate measures are not taken to control radon emissions from mill tailings piles, the public exposure to the source would exceed its exposures to all the other radiation sources associated with the uranium fuel cycle, 13 NRC 460 (1981).

The Commission recognized that localized as well as generalized exposure would occur, but it did not discuss the potential radiation doses that might result. A staff report (NUREG-0757) estimates that, at a distance of 100 meters downwind from a properly stabilized tailings pile, an individual could receive an incremental dose of 14 millirems per year above the background level. Thus, it is important to consider not only generalized but localized effects in determining whether exposures from an NRC-licensed source should be regarded as de minimis.

Nuclear Waste Management

Thus far, attention in the NRC has been principally directed to the use of de minimis risk concepts in regulating low-level rather than high-level radioactive waste management. For example, Cunningham notes the recognition of the need by the Nuclear Regulatory Commission for a de minimis approach in its regulations governing disposal of low-level radioactive waste. Such an approach would establish minimum levels of radioactivity that would have to be present before disposal in the manner prescribed for radioactive waste would be required. On December 27, 1983, the Nuclear Regulatory Commission published a final rule setting for the licensing requirements for land disposal of radioactive wastes. In the notice of rulemaking (47 FR 57446), the Commission endorsed the concept of establishing de minimis levels for radioactive wastes, characterizing such levels as those "below which there is no regulatory concern." The Commission noted that the establishment of de minimis levels would reduce the costs of radioactive waste disposal and would conserve space in disposal facilities designed for wastes having much higher activities. Thus, although it used the term *de minimis*, the Nuclear Regulatory Commission appeared to have in mind the type of cost-benefit balancing usually employed in the issuance of specific exemptions. It reiterated its belief, articulated in the notice of proposed rulemaking, that examining waste streams on a case-by-case basis would yield quicker and better results than would setting a generic limit. It recognized the desirability of a generic limit, however, and announced its plans to work toward this goal over the next few years. In order to accelerate its efforts to set standards for radioactive waste disposal by less restrictive means, the Commission provided guidelines for the submission of petitions for rulemaking seeking to establish certain waste streams as being of no regulatory concern. Finally, it noted that individual licensees may continue to request approval for alternative disposal methods pursuant to Section 20.302. That section provides a mechanism for approval of proposal mechanisms not otherwise authorized by the regulations. It remains to be seen whether the minimum levels for radioactive waste will be determined solely on the basis of a finding that the risk is de minimis or whether consideration of cost and benefit will also be taken into account.

THE ESTABLISHMENT OF CUTOFF LEVELS OF REGULATORY CONCERN IN THE TREATMENT OF SEVERE ACCIDENT ISSUES

The Need for Safety Goals and a Safety-Cost Tradeoff Criterion

The notion of establishing a de minimis risk or cutoff standard of acceptable risk is rooted in the more general issue of how safe is safe enough? The latter is also the focus of NRC's efforts in the development of safety goals.[29,30] Particularly after the accident at Three Mile Island, there arose an intensified interest in establishing more definitive safety goals for commercial nuclear power. For example, the NRC's Special Inquiry Group made the following criticism and recommendation regarding the establishment of safety goals or standards:[31]

> More surprising than that the Commission spends very little time managing or setting goals for the NRC staff is the fact that until recently it has spent very little time as a Commission deliberating or deciding any of the broad or important issues relating to reactor safety—the subject the public no doubt believes is the Commission's highest priority and certainly its raison d'etre (p. 114). As we have said elsewhere in this report, we believe the ultimate judgment of how safe is safe enough is a judgment for the Executive and Congress. The Commission should probably undertake in the first instance to articulate a proposed standard for their consideration and for public discussion, so that the value judgment about how much risk is acceptable from commercial nuclear plants can be thoroughly and publicly aired (p. 152).

The President's Commission on the Accident at Three Mile Island did not undertake to examine "how safe is safe enough" or the broader question of nuclear versus other forms of energy.[32] However, the Commission's report contains several references to "high standards for plant safety" recommended by the President's Commission (pp. 24, 64) and to NRC responsibility in providing an "acceptable level of safety" (p. 56). Also, the President's Commission noted the need for the "effective pursuit of safety goals" (p. 61). Perhaps the most explicit recommendation involving safety goals is that included in "the agency's general substantive charge should be the requirement to establish and explain safety-cost tradeoffs . . ." (p. 63).

Initiatives on behalf of safety goal development did not start from ground zero. For example, as noted by former Commissioner Joseph M. Hendrie, Appendix A to 10 CFR Part 50 sets out 64 general design criteria for quality assurance; protection against fire, missiles, and natural phenomena; limitations on the sharing of systems; and a number of other protective safety requirements.[33] In addition to the NRC regulations, there are numerous supporting guidelines that contribute importantly to the achievement of safety goals. These include regulatory guides (numbering in the hundreds); the Standard Review Plan for reactor license applications, NUREG-75/087 (17 chapters); and associated technical positions and appendices in the Standard Review Plan.

Severe Accident Policy and Safety Goal Developments

This widely felt need to reduce the risk of severe nuclear accidents, such as the one at TMI or those that would release severe fatality-producing levels of radionuclides to the general public, led to the development of the TMI Action plan.[34,35] The main driving force behind this plan was to reduce the risk of severe accidents and to provide a greater margin of assurance that the resulting level or risk was acceptably small. First issued in May 1980, Section II.B of that plan deals with the siting of plants and the requirements for coping with severe accidents. Consistent with that plan, the Commission has already issued two final rules concerning hydrogen control issues in degraded-core cooling (46 FR 58484, December 2, 1981, and 50 FR 3498, January 25, 1985). The concept of a generic rulemaking to reach final decisions on severe accidents also took form in the TMI Action Plan, Task II.B.8, "Rulemaking Proceeding on Degraded Core Accidents." This plan envisioned a long-term rulemaking extending beyond 1982 to establish policy, goals, and requirements related to accidents involving core damage greater than the present design basis for all classes of reactors: those operating, under construction, proposed for construction, or proposed as new standard plant designs.

Thus, the development of a severe accident policy was conceived at the outset as a companion effort of safety goal developments. The Proposed Commission Policy Statement on Severe Accidents and Related Views on Nuclear Regulation issued by the Commission on April 13, 1983 (48 FR 16014) provided sections on the relationships of severe accident policy to policy on safety goals and the use of probabilistic risk assessment in severe accident decision making. The final Commission Policy Statement on Severe Reactor Accidents Regarding Future Designs and Existing Plants was issued on August 8, 1985 (50 FR 32128), along with a companion document, NUREG-1070, that presented a fuller discussion of the relationships of severe accident policy to policy on safety goal development and the use of probabilistic risk assessment as a complement to traditional deterministic engineering analysis and judgment.[32]

In an interrelated set of safety goals, it appears that some of the elements should be quantitative goals and that others will be stated in relative terms. Because zero risk is an impracticality for any technological or energy alternative, a number of writers on nuclear risk point out that, to be meaningful for social decision making, comparison of the risks of energy options is an essential criterion. The Ritter bill (HR 4939) is directed toward establishing comparative risk assessments for various technologies. This concept was expressed most succinctly by Commissioner Hendrie:[33]

> My own rule of thumb for achieving the adequate protection and no unreasonable risk standards for nuclear plants—these are the phrases used, without further explanation, in the Atomic Energy Act—is that the risk to individuals should be small compared to other risks in life and that nuclear plants should present no more societal risk than other methods available for providing the bulk electricity supply (p. 12).

In May 1983 the Commission proposed for a two-year trial use period the following qualitative safety goals and quantitative design objectives (QDOs):[30]

1. Individual members of the public should be provided a level of protection from the consequences of nuclear power plant operation such that individuals bear no significant additional risk to life and health (p. 11).
2. Societal risks to life and health from nuclear power plant operation should be comparable to or less than the risks of generating electricity by viable competing technologies (particularly coal-fired plants) and should not be a significant addition to other societal risks (p. 11).
3. The risk to an average individual in the vicinity of a nuclear power plant of prompt fatalities that might result from reactor accidents should not exceed one-

tenth of 1% (0.1%) of the sum of prompt fatality risks resulting from other accidents to which members of the U.S. population are generally exposed (p. 12).

4. The risk to the population in the area near a nuclear power plant of cancer fatalities that might result from nuclear power plant operation should not exceed one-tenth of 1% (0.1%) of the sum of cancer fatality risks resulting from all other causes (p. 12).

5. The likelihood of a nuclear reactor accident that results in a large-scale core melt should normally be less than 1 in 10,000 per year of reactor operation (p. 14).

6. The benefit of an incremental reduction of societal mortality risks should be compared with the associated costs on the basis of $1000 per person-rem averted (p. 13).

Further perspective is added to the justification of these goals and their interpretation by the following statement:

> The Commission adopts this 0.1% ratio of the risks of nuclear power plant operation to the risk of mortality from non-nuclear plant origin to reflect the first qualitative goal, which would provide that individuals bear no significant additional risk. However, this does not necessarily mean that an additional risk that exceeds 0.1% would by itself constitute a significant additional risk. The 0.1% ratio to other risks is low enough to support an expectation that people living or working near nuclear power plants would have no special concern due to the plant's proximity (p. 12).

The background data on prompt (or accident) fatalities from all causes in the United States show that in 1982 there were 93,000 deaths in a population of 231 million, or a fatality rate of 4×10^{-4} per year.[15] At 0.1% of this level, the prompt fatality QDO is a risk standard of 4×10^{-7} per year. This is a standard that is substantially below the de minimis risk standard of 10^{-5} to 10^{-6} per year of exposure proposed by various policy analysts for risks that are commonly accepted with little or no concern by individuals in daily living. Also, in 1983 an estimated 440,000 persons died of cancer in the United States, or about five times the number of accident deaths. This is a rate of 1.9 cancer deaths per 1000 population. At 0.1% of this background rate, the second QDO for severe nuclear accident risk is 1.9×10^{-6} per reactor year of operation. This QDO is substantially less limiting than the prompt fatality QDO.

However, many in our society do not regard a comparison of nuclear risk with everyday risks of prompt fatalities as being appropriate. This is principally because of the catastrophic potential of nuclear risk and its involuntary imposition.[27] Regarding involuntary risk, Starr, Rudman, and Whipple note that an individual exposed to an involuntary risk is fearful of the consequences, makes risk aversion his goal, and therefore demands a level for such involuntary risk exposure many times less than would be acceptable on a voluntary basis.[36] Yet, risks imposed involuntarily on individuals may be perceived to raise little or no concern if the perceived level of direct and indirect personal benefits are substantial and the risks are so low as to be of trifling significance. In this regard, the establishment of threshold values or de minimis risk standards by certain regulatory agencies has been commended by the courts in several important decisions affecting the regulation of risks of benzene use [448 U.S. 607 (1980), pp. 607–724] and of air pollution from coal-fired electric generating plants [636 F. 2d 323–411 (1979)]. Indeed, although not formally designated as de minimis risk standards, the above

QDOs clearly possess the key attributes of such threshold standards inasmuch as the Commission adopted this 0.1% ratio of the risks of nuclear power plant operation to the risk of mortality from nonnuclear plant origin to reflect the first qualitative goal, which would provide that individuals bear *no significant additional risk*.

This implicit de minimis risk nature of NRC's fatality-risk QDOs was noted by Harvard professor Richard Wilson in a commenting letter in May 17, 1982, to the Commission regarding the proposed policy on safety goals:[37]

> All the words in NUREG-0880 *imply* that the risk levels are to be considered as *de minimis*— that is, to be accepted without any explicit consideration of benefits. I believe that this *must be* explicitly stated to avoid confusion and subsequent legal problems, and to maintain the possibility of a more flexible decision by the Commission if it should prove necessary; I believe the Commission should state that on a case by case basis, they may allow higher risks on expectation of benefit.

It is noted that the fatality risk QDOs do not address the question of whether these are appropriate surrogates for the possibly de minimis nature of the nonfatal health and other socioeconomic risks of demonstrated concern to the public.[38] A recent draft report, "The Assessment of Background Data for NRC Safety Goal Evaluation," suggests that the de minimis nature of the fatality-risk QDOs might not reflect a de minimis level of concern for nonfatal risk attributes of severe nuclear accidents.[23] If the Commission comes to adopt this view, then ALARA-type analysis of risks, costs, and benefits might be required, at least on a generic basis, to reach a regulatory policy or rulemaking on the acceptable safety of a given nuclear plant design and its operational risk management practices. This would not negate, however, the useful perspective of the explicit recognition that the above QDOs, if met, would constitute an appropriate level of de minimis risk from the standpoint of fatality risk of individuals living or working in the vicinity of a nuclear power plant.

On August 4, 1986, the Nuclear Regulatory Commission published in the *Federal Register* its approved "Policy Statement on Safety Goals for the Operation of Nuclear Power Plants" (51 FR 28044). This policy statement leaves unchanged the two proposed qualitative safety goals and the two quantitative design objectives (see items 1–4 above). It deletes the plant performance objective that the likelihood of a large-scale core meltdown should normally be less than 1 in 10,000 per year of reactor operation (see item 5). It also deletes the cost-benefit guideline (see item 6) that the benefit of an incremental reduction of societal mortality risks should be compared with the associated costs on the basis of $1000 per person-rem averted.

Depending on weather conditions at the time of the accident and other factors, many (but not all) persons residing two or more miles from the plant would receive radiation doses so low as to constitute a de minimis level of cancer fatality risk. The proportion of persons receiving de minimis risk levels would generally be expected to increase as the distance from the plant site increases. Although not explicitly stated in the safety goal policy statement, these de minimis risk implications are inherent in the Commission's decision to reduce the distance for computing the average cancer fatality risk to 10 miles from the 50-mile distance stated in the 1983 proposed policy statement. This new policy statement provides the following simplified explanation of this change:

> The change to 10 miles could be viewed to provide additional protection to individuals in the vicinity of the plant, although analyses indicate that this objective for cancer fatality will not be

the controlling one. It also provides more representative societal protection, since the risk to the people beyond 10 miles will be less than the risk to the people within 10 miles.

In the policy statement, the Commission recognizes that the staff will require specific guidelines to use as a basis for determining whether a level of safety ascribed to a plant is consistent with the safety goal policy. As a separate matter, the Commission intends to review and approve guidance to the staff regarding such determinations. It is currently envisioned that this guidance would address matters such as plant performance guidelines, indicators for operational performance, and guidelines for conduct of cost-benefit analyses. This guidance would be derived from additional studies conducted by the staff and resulting in recommendations to the Commission. The guidance would be based on the following general performance guideline which is proposed by the Commission for further staff examination:

> Consistent with the traditional defense-in-depth approach and the accident mitigation philosophy requiring reliable performance of containment systems, the overall mean frequency of a large release of radioactive materials to the environment from a reactor accident should be less than 1 in 1,000,000 per year of reactor operation.

In keeping with the common-use concept of de minimis risk explored in this chapter, a case can be made that this proposed performance guideline provides for a de minimis level of risk for *all* off-site persons receiving *any* radiation from a severe nuclear accident. Even if a person were to receive a sufficient radiation dosage to "guarantee" prompt or delayed cancer fatality, the probability of such a tragic event is at most 1 chance per million plant-years of operation. Moreover, the risk of off-site property damage would also be subject to this same low-probability coefficient. Nevertheless, the Commission's policy statement recognizes that apart from their health and safety consequences, severe core damage accidents can erode public confidence in the safety of nuclear power and can lead to further instability and unpredictability for the industry. In order to avoid these adverse consequences, the Commission intends to continue to pursue a regulatory program that has as its objective providing reasonable assurance, while giving appropriate consideration to the uncertainties involved, that a severe core damage accident will not occur at a U.S. nuclear power plant.

The Commission further notes that these safety goals and implementation guidelines are not meant as a substitute for NRC's regulations and do not relieve nuclear power plant permittees and licensees from complying with regulations. To provide adequate protection of the public health and safety, current NRC regulations require conservatism in design, construction, testing, operation, and maintenance of nuclear power plants. A defense-in-depth approach has been mandated in order to prevent accidents from happening and to mitigate their consequences. Siting in less populated areas is emphasized. Furthermore, emergency response capabilities are mandated to provide additional defense-in-depth protection to the surrounding population.

CONCLUSIONS

This summary review of NRC's past and current usage and interest in the de minimis risk concept revealed a number of examples of explicit reference to this concept as well as other discussions of NRC's regulatory decision-making regarding radiological risk levels falling within the implicit range of common-use values of de minimis risk. These include (1) the proposed revision to 10 CFR Part 20 regulations governing radiological protection that is currently before the Commission, (2) the use of the ALARA (as low as reasonably achievable) concept in decisions setting regulatory exemption levels that may fall below or above the level of a common-use concept of de minimis risk, (3) explicit attention to the de minimis risk concept in setting acceptable surface contamination levels in nuclear facility decommissioning policy, (4) the implicit nature of de minimis risk concepts in using ALARA principles in developing new regulations for uranium mines and mills to increase radiological protection for the public, (5) the Commission endorsement of the concept of establishing de minimis levels for radioactive wastes, and (6) the implicit nature of de minimis risk concepts in the Commission's safety goal policy wherein quantitative design objectives are set at 0.1% of the U.S. background levels of prompt fatalities and latent cancer fatalities from all causes.

However, there is a clear need for a more coherent and integrated policy toward de minimis risk concepts and the related use of ALARA principles in regulatory decision making. There is also a need for attention to the policy consideration of whether de minimis risk concepts should be focused on an individual risk perspective and/or one involving the aggregate of societal risk possibly over many generations into the future.

REFERENCES

1. R. H. Clarke, "Radiological Protection Aspects of Exemption Levels in the Nuclear Fuel Cycle," Seminar on *Interface Questions in Nuclear Health and Safety,* sponsored by the OECD Nuclear Energy Agency, Paris (April 16–18, 1985).
2. Richard Wilson, "Commentary: Risks and Their Acceptability," *Science, Technology, and Human Values,* 9, 2, 11–22 (Spring 1984).
3. R. E. Baker, W. S. Cool, and W. A. Mills, "NRC Draft Revision of 10 CFR Part 20. Cut-off Level for Regulatory Concern (De Minimis)," p. 17, (1983).
4. R. H. Clarke and A. Fleishman, "The Establishment of De Minimis Radioactive Wastes," IRPA 6th Congress, West Berlin (May 1984).
5. C. B. Meinhold, "Criteria for a De Minimis Level," Symposium of the U.S. Health Physics Society, New Orleans (June 3–8, 1984).
6. ICRP, *Radiation Protection: Recommendations of the International Commission on Radiological Protection* (New York: Pergamon Press, 1977), ICRP Publication No. 26.
7. D. Beninson and B. Lindell, "Bases and Trends in Radiation Protection Policy," Seminar on *Interface Questions in Nuclear Health and Safety,* sponsored by the OECD Nuclear Energy Agency, Paris (April 16–18, 1985).
8. Leonard Bickwit, Jr., "Adequate Protection of the Health and Safety of the Public," Memorandum to the Commissioners from the NRC Office of General Counsel (October 18, 1979).
9. USNRC, *Press Release No. 75-108* (April 30, 1975).
10. USNRC, *Programmatic Environmental Impact Statement related to decontamination and disposal of radioactive wastes resulting from March 28, 1979 accident Three Mile Island and Nuclear Station, Unit 2,* NUREG-0683, Supplement No. 1 (October 1984, p. 3.6).
11. EPA, *Environmental Radiation Protection Requirements for Normal Operations of Activities in the Uranium Fuel Cycle,* Environmental Impact Statement for establishing 40 CFR 190, EPA 520/4-76-016, Vol. 1 (November 1976).
12. USNRC, "Proposed Revision of 10 CFR Part 20, 'Standards for Protection Against Radiation,' " SECY-85-147-Part I (April 22, 1985).

13. M. B. Spangler, "The Need for De Minimis Risk Standards in Regulatory Decisionmaking: An Individual or a Societal Risk Concept?" in *Environmental Health Risks: Assessment and Management*, R. S. McColl, ed. (University of Waterloo Press, Ontario, Canada 1987).

14. Robert B. Minogue, Memorandum for Commissioner Asselstine on "Relationship Between Proposed Revision to 10 CFR Part 20 and EPA Radiation Protection Standards" (August 26, 1985).

15. Guy H. Cunningham, III, "The De Minimis Concept in Radiation Protection," A paper presented at the Annual Meeting of Nuclear Safety Research Association of Japan, Tokyo (June 16, 1983).

16. USNRC, *Plan for Reevaluation of NRC Policy on Decommissioning of Nuclear Facilities*, NUREG-0436, Revision 1 (December 1978).

17. USAEC, "Termination of Operating Licenses for Nuclear Reactors," Regulatory Guide 1.86, U.S. Atomic Energy Commission (June 1974).

18. K. S. Crump, "Statistical Aspects of Linear Extrapolation," Chapter 29 in *Health Risk Analysis*, C. R. Richmond, P. J. Walsh, and E. D. Copenhaver, eds. (Gatlinberg, Tenn., October 1980), 381–392.

19. National Academy of Sciences, *The Effects on Populations of Exposure to Low Levels of Ionizing Radiation, Report of the Committee on the Biological Effects of Ionizing Radiation*, BEIR III, National Academy of Sciences (July 1980).

20. Otto White, Jr. and J. E. Brower, eds., *Workshop on Problem Areas Associated with Developing Carcinogen Guidelines*, BNL 51779, Center for Assessment of Chemical and Physical Hazards, Brookhaven National Laboratory, Upton, N. Y. (June 1984).

21. A. B. Fleishman, "The Significance of Small Doses of Radiation to Members of the Public," National Radiological Protection Board, Chilton, Didcot, Oxon, United Kingdom (March 1985).

22. USNRC, *Final Environmental Impact Statement on Uranium Mining*, NUREG-0706, 3 Vols., U.S. Nuclear Regulatory Commission (September 1980).

23. Miller B. Spangler, "The Assessment of Background Data for NRC Safety Goal Evaluation," Draft report prepared for the U.S. Nuclear Regulatory Commission (April 24, 1985).

24. H. T. Peterson, Jr., "Regulatory Implications of Radiation Dose–Effect Relationships," *Health Physics* 47(3): 345–359, (September 1984).

25. H. Kato and W. J. Schull, "Studies of the Mortality of A-Bomb Survivors, Part I., Cancer Mortality," *Radiation Research 90*, 395–432 (1982).

26. D. MacLean and P. Brown, eds., *Energy and the Future*, Rowman and Littlefield, (Totowa, N. J., 1983).

27. M. B. Spangler, "An International Perspective on Risk and Equity Issues Associated with the Coal and Nuclear Fuel Options," *Journal of Public and International Affairs 5* (1): 101–121, (Winter 1984).

28. M. B. Spangler, "Policy Issues Related to Worst Case Risk Analyses and the Establishment of Acceptable Standards of De Minimis Risk," in *Uncertainties in Risk Assessment and Management*, V. Covello, A. Moghissi, and V. T. T. Uppuluri, eds. (Plenum, N. Y. 1987).

29. R. J. Mattson et al., "Concepts, Problems, and Issues in Developing Safety Goals and Objectives for Commercial Nuclear Power," *Nuclear Safety 21*(6): 703–716, (November–December, 1980).

30. USNRC, *Safety Goals for Nuclear Power Plant Operations*, NUREG-0880, Rev. 1 (May 1983).

31. M. Rogovin et al., *Three Mile Island: A Report to the Commissioners and to the Public*, NUREG/CR-1250, Nuclear Regulatory Commission (January 1980).

32. J. G. Kemeny et al., *Report of the President's Commission on the Accident at Three Mile Island: The Need for Change: The legacy of TMI*, GPO (October 1979).

33. J. M. Hendrie, "Nuclear Safety and the Regulation of Nuclear Technology," Remarks by NRC Commissioner Hendrie before the *First Texas Symposium on Energy*, The University of Texas at Dallas, Richardson, Tex. (July 9, 1980). Available from NRC Public Document Room as S-12-80.

34. USNRC, *NRC Action Plan Developed as a Result of the TMI-2 Accident*, NUREG-0660, Vols, 1 and 2, Rev. 1 (August 1980).

35. USNRC, *Clarification of the TMI Action Plan Requirements*, NUREG-0737 (November 1980).

36. C. Starr, R. Rudman, and C. Whipple, "Philosophical Basis for Risk Analysis," in *Annual Review of Energy 1*, J. M. Hollander and M. K. Simmons, eds., Annual Reviews, Inc., (Palo Alto, Calif, 1976).

37. Richard Wilson, Letter to Samuel Chilk, Secretary, U.S. Nuclear Regulatory Commission, "RE: Proposed policy on safety goals for nuclear power plants (45 FR 71023)," dated May 17, 1982.

38. P. Slovic, B. Fischhoff, and S. Lichtenstein, "Rating the Risks," *Environment 21*, (3): 14–39, (April 1979).

13

The Feasibility of Establishing a De Minimis Level of Radiation Dose and a Regulatory Cutoff Policy for Nuclear Regulation

Joyce P. Davis

INTRODUCTION

Scope of Study

This report discusses the feasibility of a de minimis policy for radiation exposure guidance and regulation. This policy would establish a threshold below which regulation of radiation sources, practices, or exposures would be deliberately and specifically curtailed. While a similar policy would probably be appropriate for other environmental agents (e.g., chemical carcinogens), discussion of that aspect of the subject is outside the scope of this report.

The policy discussed in this chapter would be applicable to the regulation of all radiation exposure, whether by the Nuclear Regulatory Commission (NRC), Environmental Protection Agency (EPA), Food and Drug Administration, or state agencies. The focus of this report, however, is on the regulation of NRC licensees, particularly power reactor licensees.

In the absence of such a policy, it is likely that individual practical cutoff levels for different applications will continue to be set by various regulatory bodies, with little consistency in the methodologies used to establish them, and without acknowledgment that such cutoffs are a reasonable approach to radiation regulation.

Joyce P. Davis • General Physics Corporation, Columbia, Maryland 21044.

The De Minimis Concept

The law has long recognized that there are trivial matters that need not concern it. The maxim *de minimis non curat lex*, "the law does not concern itself with trifles," expresses that principle.* As a practical matter, in the area of administrative law, many regulations and regulatory practices involve explicit cutoff levels, below which the regulator does not concern himself because the harm involved in so doing is considered negligible. Some of these are discussed later in this chapter. Of course, the applicability of such cutoff levels depends on the nature of the matter regulated and the statute under which the regulation is carried out.

Recently, several experts in the field of radiological health have suggested that a level of radiation exposure be identified as being small enough to be considered negligible. This has been variously termed a de minimis dose,[1] a practical threshold,[2] and a level safe enough to exempt,[3] among others.

It is reasonable to make a distinction between doses of radiation that scientists consider negligible and those that regulators, acting for the public, determine to be of no regulatory concern. Ideally, these two levels should coincide. However, in practice, since their bases must differ, they may not be identical. At the very least, however, the scientific de minimis level should be an important input to the regulatory process.

For the purposes of this report, which is concerned with regulation of exposure to ionizing radiation and radioactive materials, terms are defined as follows:

De Minimis Level: A value of increment in dose equivalent or radioactive material concentration, above background, that would be deemed trivial, or of no concern in decision making from the point of view of his own or his family members' individual risk of harm, by an expert in the field of radiological health.

Regulatory Cutoff Level: A value of radiation dose equivalent, radioactive material concentration, or other related quantity, at or below which there is not further regulatory concern.

In this chapter, *exposure* is used in the general sense and not as a defined scientific quantity. *Dose* is used as a shorthand synonym for "dose equivalent." *Dose equivalent*, in units of rems or millirems (mrems), measures the biological effect of radiation absorbed by the human body. "Collective dose equivalent" or "population dose" is the summation of individual dose equivalents over a population group and is expressed in units of person-rems. *Risk* is used in an objective sense, defined as a function of the probability of deleterious effects. Except where specifically noted, it is not used in the subjective sense (i.e., "perceived risk").

---

* *De minimis non curat lex*, The law does not care for, or take notice of, very small or trifling matters.
"The law does not concern itself about trifles" -Cro. Eliz 353.
"Thus, error in calculation of a fractional part of a penny will not be regarded" - Hob 88.
"So, the law will not, in general, notice the fraction of a day" - Broom, Max 142. (From Black's Law Dictionary, 4th edition, 1951)
Some of the recent cases discussing the application of the de minimis concept in environmental law are presented in "Risk and Benefit in Environmental Law" by P. F. Ricci and L. S. Molton (214 *Science*, p. 1096, December 4, 1981).

In the following two sections, this report will examine whether a de minimis level (or levels) can be established for radiation exposure (The Technical Context), and whether a regulatory cutoff level would be feasible (The Legal Context). Further sections examine the development of a regulatory cutoff policy and the associated problems and benefits. The final section presents the conclusions and recommendations of the study.

THE TECHNICAL CONTEXT

The Biological Effects of Radiation at Low Doses and Low Dose Rates

This report is concerned with levels of average radiation dose rate that are at or below the current limits for exposures to workers (average 5 rems per year) and the general public (500 mrems per year). This is within the range of "low dose" for single exposure, defined by the National Council on Radiation Protection and Measurements (NCRP) as generally 0 to 20 rems, and "low dose rates" defined as dose rates of the order of 5 rems per year or less, a rate that includes those for background radiation exposure, and radiation guides for workers and for the general public. Figure 1 shows the relative levels of radiation exposure.

Observed Health Effects. No clinical ill-health effects proven to be causally related to exposure have even been observed in human individuals exposed to low doses or low dose rates of ionizing radiation. At high doses and dose rates, by contrast, an acute radiation syndrome is seen, teratogenic effects can occur, and the incidence of cancer in a exposed population is observed to be increased above the natural incidence. No genetic effects of radiation have been observed in any human population at any dose, but the likelihood of such effects is indicated by animal studies. In the sections that follow, the emphasis is on the carcinogenic effects of radiation. Most radiobiologists agree that the risk of genetic effects is no greater than the cancer risk, and therefore the use of cancer risk alone in estimating risk does not introduce gross error.

Effects like the acute radiation syndrome and teratogenic effects are defined as "nonstochastic" effects. Their severity varies with dose, and a threshold exists below which no effect is observed[4] (see Figure 2). At low doses and dose rates of ionizing radiation, no nonstochastic health effects have been observed, and none is believed to be of importance.

Another class of effects has been defined as "stochastic effects," which are those for which the *probability* of an effect occurring, rather than its *severity*, is regarded as a function of dose. Many radiobiologists believe that cancer induction and genetic effects, which are observed to be *stochastic* at high doses, are effects that might be caused by radiation at low doses or dose rates, down to doses approaching zero (i.e., that there is no threshold for the probability of induction of such effects, see Figure 2). It should be noted that for the stochastic effects of radiation, even at quite high doses, most exposed individuals do not exhibit the effect. For example, at a dose of about 100,000 mrems, no more than a few percent of the irradiated individuals would be expected to develop a fatal cancer due to the exposure. Such data are generally expressed in terms of "risk" of effect—i.e., the probability of an individual effect being observed, or the number of

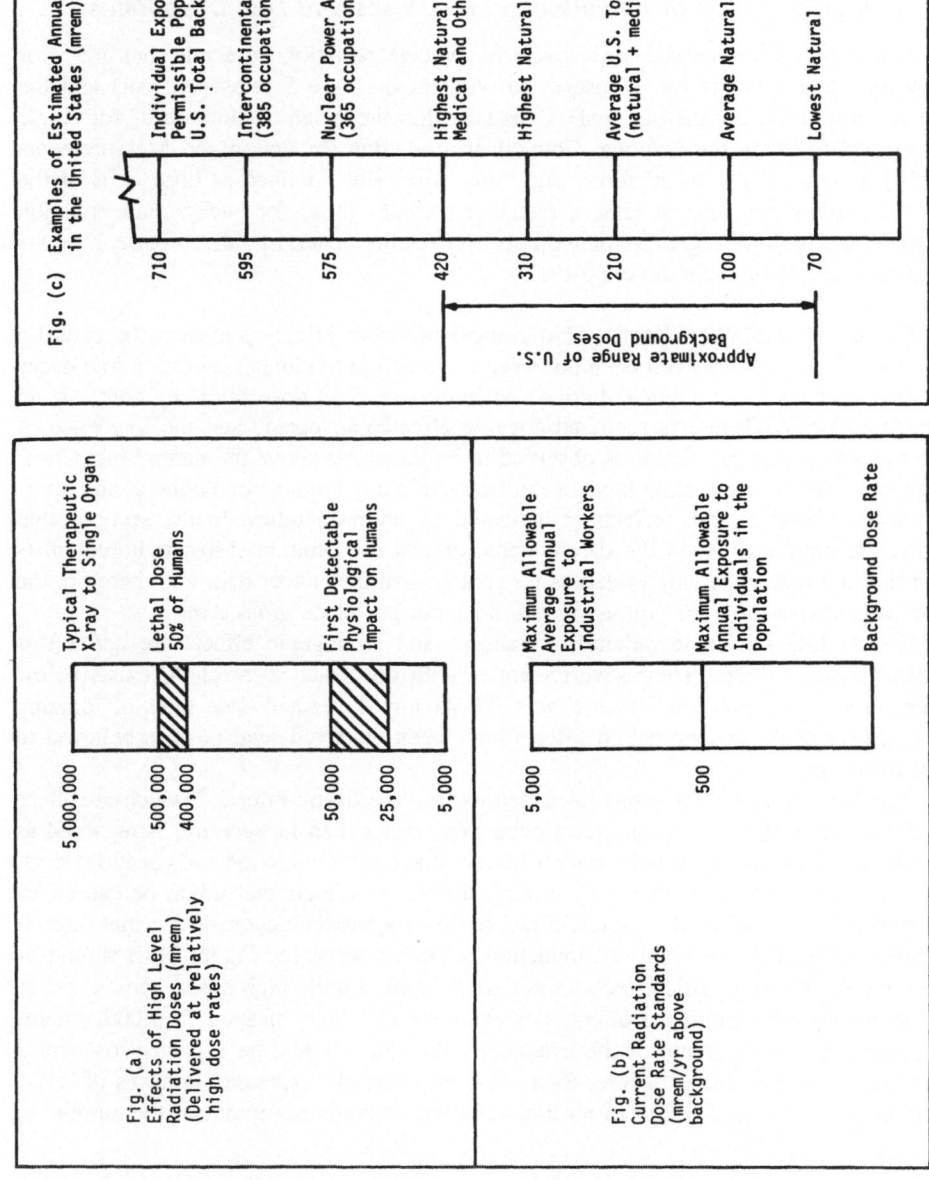

Figure 1. Relative levels of radiation exposure.

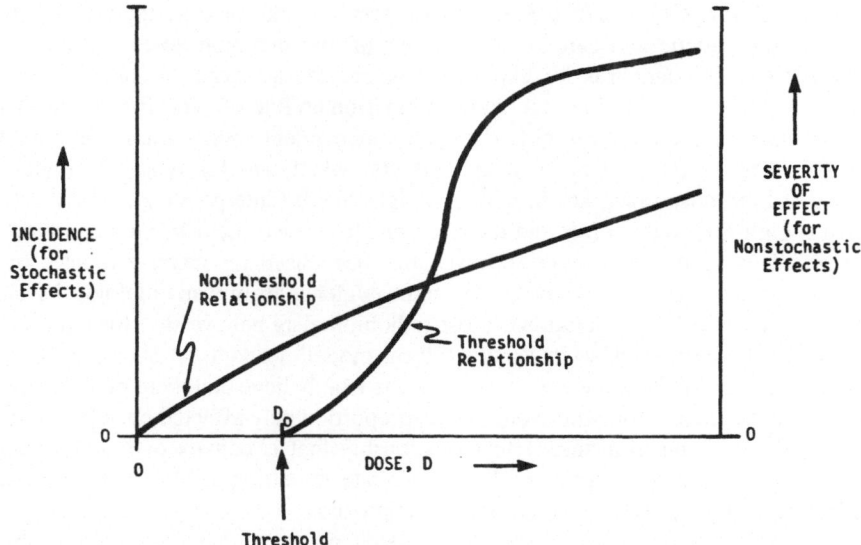

Figure 2. Examples of threshold and nonthreshold dose–effect relationships.

affected individuals to be observed in a large population, say 100,000 or one million people.

Because the probability of a radiation-induced cancer, if nonzero, would be very small at low dose rates, and because radiation-induced cancers are clinically indistinguishable from those occurring naturally (or caused by other carcinogenic agents), the only practical way to detect such a radiation effect is by means of a large-scale statistical study. It is theoretically possible to detect such effects, if they exist, by epidemiological studies of comparable "exposed" and "unexposed" populations. If an excess of cancers above the expected natural incidence is detected in the exposed group, the inference that the increase is causally related to the exposure may be made.

As a practical matter, however, such epidemiological studies are very difficult. In order to detect these effects, which would be quite small compared to natural incidence, and even to variations in natural incidence over space, time, environmental variables, and statistical uncertainties, very large populations must be included in the studies, and the effects of confounding factors like smoking habits must be minimized.[5] Because of this, the National Academy of Sciences' Committee on the Biological Effects of Ionizing Radiation (BEIR) has stated that "it is unlikely that the carcinogenic or teratogenic effects of doses of low-LET radiation administered at a dose rate of about 100 mrems per year or less will be demonstrable in the foreseeable future. For higher (low) dose rates, e.g., a few rems per year over a long period, a discernable carcinogenic effect could become manifest." The method of demonstration that the BEIR committee had in mind is the epidemiological study.

As discussed below, the epidemiological evidence is inconclusive, and, according to some observers, may tend not to support the *no threshold* hypothesis.[6-13,83] It may be that the so-called stochastic effects become nonstochastic at low doses and dose rates.

Inferred and Extrapolated Health Effects. Confronted by the lack of knowledge about the genetic and carcinogenic effects of low radiation doses and dose rates, and the need to set standards for exposure that provide a margin of safety, standards-setting bodies [e.g., NCRP, International Commission on Radiological Protection (ICRP)] developed the linear, nonthreshold, dose-rate-independent model for radiation effects prediction. The available data on health effects, which are for relatively high doses delivered at high dose rates, are fit with a straight line by interpolating linearly between the lowest high-dose data points and the point of zero excess incidence at zero dose[4] (see Figure 3). This relatively simple approach has been used to obtain estimates of risk (probability of an effect) at low doses. Given the evidence from many biological systems, as well as theoretical considerations, most radiobiologists believe that for exposures to radiation at low doses or dose rates, this "linear model" leads to an overestimate of the risk in most cases. While there are a few scientists who believe that there may be sensitive subgroups in the population who would be disproportionately affected even by the lowest of doses,[14,15] resulting in a dose–effect relationship that is convex over some range of doses, the use of the "conservative" linear estimate in setting safety limits is generally considered a prudent, and not unreasonable, approach.

The linear, nonthreshold, dose-rate-independent model has also been applied to estimating the risk of radiation exposure, for comparison with other risks, as a basis for decision making. However, in this application, the fact that the risk estimate is an "upper-limit" must be considered. When radiation risks are compared to other actual risks, or to actual benefits and costs, the use of extrapolated "upper-limit" values is misleading. The NCRP has cautioned against the use of "conservatively" estimated radiation risk values for comparison with "realistically" estimated risks of some alternative technologies. Most regulatory bodies, however, have continued to use the linear nonthreshold dose-

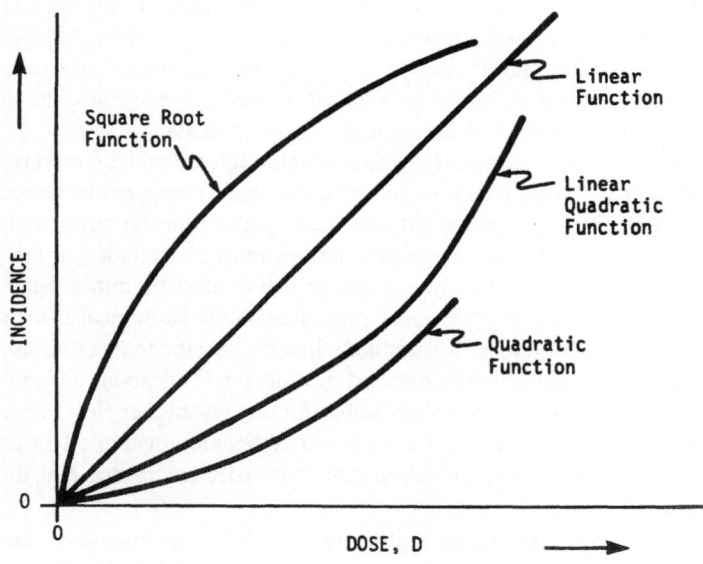

Figure 3. Examples of nonthreshold stochastic dose–effect relationships.

independent hypothesis to quantitatively estimate risks, often presenting such estimates as though they were actual expected values.

Efforts are being made to develop methods for estimating radiation risks that are more realistic. They are, of course, limited by the lack of data on very low dose and dose rate effects, and by the unlikelihood that such data will become available in the foreseeable future. The simplifying assumptions of the conservative model, and more realistic approaches, are discussed below:

Linearity. In its latest report, the BEIR Committee majority adopted a curvilinear relationship as the best estimate of the dose–effect curve.[16] They selected a linear-quadratic form (see Figure 3). One dissenting member of the committee endorsed the pure linear relationship and another one supported a pure quadratic.* The linear-quadratic relationship falls between those two curves (see Figure 3). In the 1981 BEIR report, this model was applied to estimating the health effects (risks) of a single dose of 10 rems and of a lifetime exposure to a dose rate of 1 rem per year. While the BEIR report does not explicitly sanction such use, the risks associated with those levels have been extrapolated linearly to estimate risks at much lower doses and dose rates.[18]

In the linear-quadratic model, endorsed by BEIR, the dose–effect curve is linear in the very low dose range. The slope of that linear segment, however, is significantly smaller than that of the pure linear case. Thus, the estimated detriment per unit dose is lower, in this lowest dose range, than it would be were the pure linear model used. At the lowest doses it is unlikely that the *true* shape of the curve will ever be known; however, at these levels of dose the predicted risks are quite "small," under almost any definition of that word, regardless of the shape of the curve.

Nonthreshold

Absolute Threshold. Many experts believe it likely, in light of current knowledge, that there is no absolute threshold for carcinogenic and genetic effects and risks. In other words, at no dose, however small, can it be said as a scientific certainty that there is no risk. As noted above, however, the risks at very low levels of dose are so small that, although finite, they are close to zero. For conservatism in regulatory applications, it has generally been assumed that there is no absolute threshold for stochastic effects.

Practical Threshold of Latency. There may not be an absolute threshold for carcinogenic or genetic effects. Indeed, it may be possible, eventually, to prove on theoretical grounds that a threshold does not exist, although that proof might not be experimentally

---

* In the BEIR report[16] there appears, in addition to the committee's chapter entitled "Somatic Effects: Cancer," two statements by dissenting scientists. Dr. Edward P. Radford, M.D., stated his opinion that the BEIR Committee's earlier decision to reaffirm the applicability of the linear no-threshold dose–response relationship was the correct one, and that such a decision was not so conservative as had been thought at the time of the BEIR I report. While an underestimate of risk would be unlikely for low-LET (beta and gamma) radiation, Dr. Radford stated that the view that the linear extrapolation greatly overestimated the risk at low doses appeared unwarranted to him.[15]

Dr. Harald H. Rossi also provided a separate statement. He expressed his belief that the BEIR report is deficient because many of its risk estimates are still based on the "linear hypothesis," despite mounting contrary evidence, and because it fails to present explicitly data that indicate risk factors for low-LET radiation less than the lowest given in the report.[17]

verifiable. However, there is evidence for a "practical threshold," at least for some effects of radiation.

Dr. Robley Evans[19] noted that for radiogenic tumors in several species, caused by skeletal doses from radium-226 radiation, the data suggest that tumor appearance time ("latency period") increases as the dose decreases. If this is so, then there is a domain of small dosage for which the tumor appearance time exceeds the maximum life-span of individuals of that species. At doses in this domain, the radiogenic tumors would appear in an irradiated population with negligible frequency. Dr. Evans states that this concept of "practical threshold dose," below which the incidence of a radiogenic effect is extremely low during the remaining survival time, is "an old familiar one in human radiobiology."[19] He suggests this hypothesis to explain the threshold observed in radiogenic tumor incidence (see Figure 4).

Practical Threshold of Integer Health Effects in a Population. Another type of practical threshold may be observed when the risk to a population, rather than to an individual, is considered. At some point the risk, estimated using a dose–effect model, becomes so low that the estimate of the number of people affected in that entire population becomes less than one. It may be reasonable to round this value off to zero. At this dose level it can be said that not even one person would be expected to be affected. The dose at which this *threshold* appears depends, of course, on the size of the population at risk as well as the dose–effect relationship assumed.

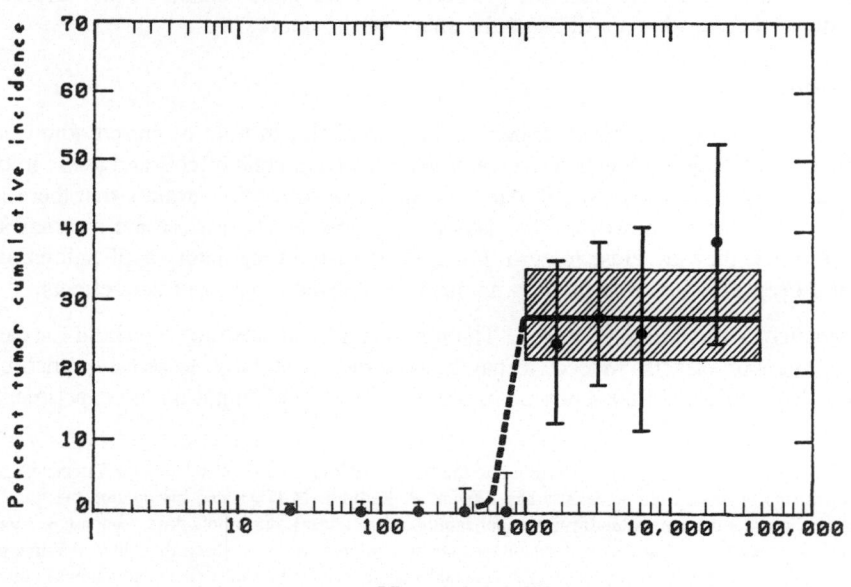

Figure 4. Practical threshold of latency. Note: This figure plots the observed radiogenic tumor cumulative incidence or occurrence in the "epidemiologically suitable" cases summarized in Dr. Robley Evens's paper, "Radium in Man" (*Health Physics* 27, pp 497–510, 11/74). The shaded region corresponds to the mean occurrence 28 +/− 6% between 1000 and 50,000 rads.

Threshold of Competing Beneficial Effects. A different type of threshold may be postulated to occur if one accepts the possibility of radiation "hormesis"—the hypothesis that low doses of radiation are actually beneficial to health, in contrast to the detrimental higher levels. Dr. Victor Bond cites several "well-done radiation studies on small mammals" indicating that exposure to long-term, low-level radiation, of the order of scores of rem delivered over months, may extend the animals' average life-span by a small amount. He states that presumably small doses "may stimulate repair, recovery and other defense mechanisms that can counter-balance small increments of injury."[6]

Dr. L. A. Sagan, in a recent article, discussed the idea of hormesis and the possibility that small doses might produce effects qualitatively different from those observed at higher doses. Many environmental agents are known to have different effects at low and at high doses. For example, "many of the essential human nutritional elements are beneficial, even necessary, to health at low doses, but are toxic or poisonous at high doses."[9]

As Dr. Sagan notes, the existence of a beneficial radiation effect does not necessarily exclude the linear theory of radiation carcinogenesis. There might be a beneficial effect at low doses superimposed on the detriment. At very low doses the total effect might be zero or even beneficial (see Figure 5). In a recent monograph, Dr. T. D. Luckey has assembled extensive literature supporting the idea of radiation hormesis.[7,12] It includes

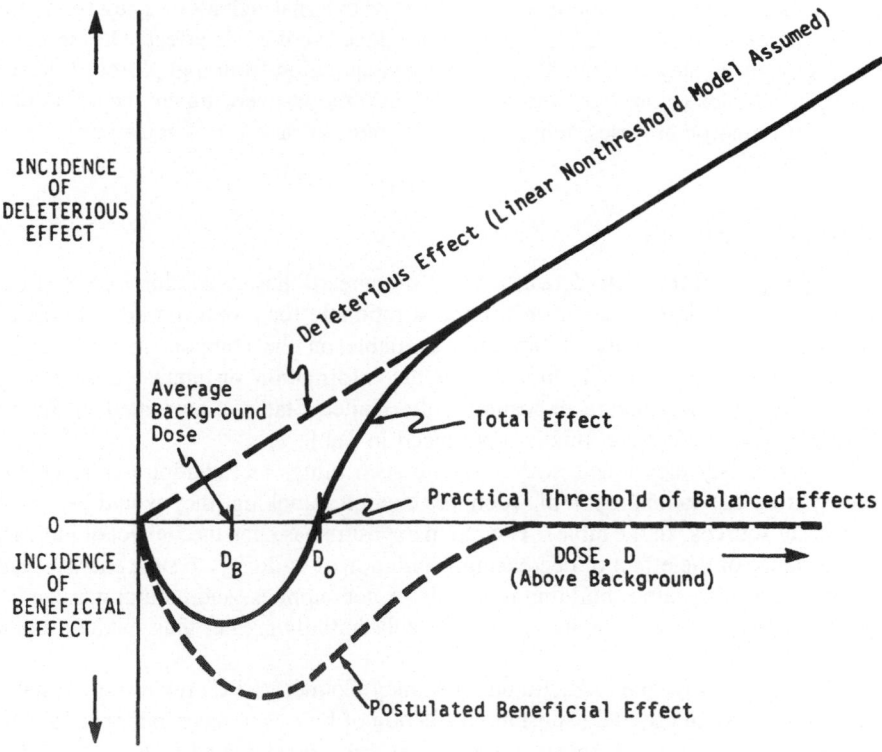

Figure 5. Threshold of competing beneficial effects.

references to apparent *reduced* carcinogenesis in human populations living in high natural radiation areas, as suggested by the work of Dr. Frigerio of Argonne National Laboratory.[20] Indeed, Dr. Luckey even presents evidence for the theory that radiation is essential for life. He believes there is an optimum level of radiation exposure somewhat higher than average background radiation dose rates.[7,12]

Proof of beneficial effects of low doses of radiation in human populations is as difficult as proof of detrimental effects, as discussed earlier. The one certainty is that the risks or benefits, whatever their nature, are very small. If a radiohormetic effect exists and could be quantified, then a *threshold* for negative effects could be established as the dose at which positive and negative effects are equal. At lower doses there might be a net benefit of radiation exposure (see Figure 5). Of course, if the beneficial effect is qualitatively different from the detrimental one, it may be difficult in practice to trade them off against one another, even if both could be expressed in quantitative terms.

Dose-rate-independent. The BEIR committee did not explicitly model a dose rate effect in its estimate of radiation risks. They stated, however, "the Committee does not know whether dose rates of gamma or X-rays of about 100 (mrems) per year are detrimental to man. Any somatic effects at these dose rates would be masked by environmental or other factors."[16]

NCRP in its Report No. 64 considered the influence of dose and its distribution over time (dose rate) on dose–response relationships for external radiation. They found that a reduction in dose rate, in general, results in a reduced biological effect. On the basis of available evidence, they estimated[4] that linear extrapolation from high doses (150 to 350 rems) and dose rates (more than 5 rems per minute) might overestimate the effect of low doses (0–20 rems) or of doses delivered at dose rates at or below 5 rems per year, by a factor of 2 to 10.

Background Radiation

Background Radiation Exposure. Life on earth has evolved in a field of background radiation. This fact has been used as a rationale for proposing de minimis dose levels. There is a large body of literature available on the components and levels of background radiation. A good summary of the information on annual dose rates for significant sources of radiation exposure in the United States is presented in the 1981 BEIR report.[16] For reference, this is reproduced in Table 1.

Not specifically mentioned in this list are such things as radiation exposure to the respiratory tract due to inhalation of radon and cigarette smoking (they would be included with "internal sources" in the table). The radon exposure has been the subject of increasing concern because of the effect of decreasing ventilation of buildings to save energy. Radon emanating from the ground, building materials, water supplies, and natural gas can build up to concentrations in interior spaces that are substantially greater than outdoor ambient levels.

In its recent study, the U.S. Radiation Policy Council (RPC) reviewed the data on indoor radon levels and found a substantial fraction of houses to have indoor radon levels more than five times the previously presumed average indoor level.[21] As seen in Table 2 (which is taken from the RPC report), the annual effective collective background whole-body dose from interior radon is about 20% of the total background dose. Recent data indicate that indoor radon levels may be much higher in some houses, and the trend is toward increasing levels because of energy conservation measures.

Sources of background exposures are often categorized into "natural" and "man-made." The distinction is not always sharp, and a new category, "man-enhanced" (sometimes called "technologically enhanced") has come into use. Cosmic rays are clearly "natural" and medical X rays are "man-made." Phosphate mine tailing piles are, in this context, "man-enhanced." But what about indoor radon? Is it "natural" if one lives in a natural cave and "man-enhanced" if one lives in a stone house? The distinction may become important if a de minimis level of radiation is defined with reference to "background" or "natural background." The definition is not complete unless the components to be considered as "background" are identified.

Variation in levels of background components over space and time have been documented. For example, Table 3 shows natural background levels (not including internal doses) in various parts of the world.[22]

Statistical measures of the variation of some data on background levels have been evaluated. For instance, Weinberg and Adler (see Appendix) have calculated the standard deviation of the distribution of average natural background radiation by states in the United States. This is shown in Figure 6.

There are also uncertainties in estimates of individual and population exposures to background radiation. These uncertainties are related to the dose measurements, and to the physical and sociological models used in making the estimates of exposure for an actual population.

Background Radiation Risks. As discussed earlier, no detrimental effects of radiation at the low dose levels associated with background have been observed. Studies have been carried out, for various populations, in areas of the world where relatively high natural background levels exist (the annual dose to certain populations in India, for example, is about 1300 mrems) without any definitive results. This is not surprising, since even if the upper-limit linear extrapolation is used, the risk of cancer induced by exposure to radiation at the levels involved is very small. For example, for the average U.S. natural background of about 100 mrems/year, the annual cancer fatality risk would be 10 in 1 million (or about 2500 cases). Such an increment is too small to be detected in the observed annual cancer fatality rate of about 2 in 1000 (or about 500,000 cases) in the U.S. population.[23]

There is no convincing evidence that background radiation has any deleterious effect at all, and it might even have a beneficial effect (see p. 153). Some of the epidemiological evidence seems to point in that direction.[11,13,20] Recent studies by Hickey et al. suggest that health effects of background radiation, if any, do not exceed (in magnitude) those of environmental chemicals, and that models implying important long-term deleterious effects of low-level radiation on humans may be invalid.[13] Furthermore, it is unlikely that a more definitive statement about background risk will be possible in the foreseeable future. In any event, most experts agree that the risks of exposure to background radiation are less than those calculated by the linear hypothesis and thus, by most measures, can be considered "small."

Background risks taken out of context, can look large, even enormous, when pop-

Table 1. BEIR Report Summary of Radiation Exposure[a]

	Exposed group				Average dose rate, mrems/yr	
Source	Description	N exposed		Body portion exposed	Exposed group	Prorated over total population
Natural background						
Cosmic radiation	Total population	220×10^6		Whole body	28	28
Terrestrial radiation	Total population	220×10^6		Whole body	26	26
Internal sources	Total population	220×10^6		Gonads	28	28
				Bone marrow	24	24
Medical X rays						
Medical diagnosis	Adult patients	105×10^6/yr		Bone marrow	103	77
Medical personnel	Occupational	195,000		Whole body	300–350[b]	0.3
Dental diagnosis	Adult patients	105×10^6/yr		Bone marrow	3	1.4
Dental personnel	Occupational	171,000		Whole body	50–125[b]	0.05
Radiopharmaceuticals						
Medical diagnosis	Patients	10×10^6 to 12×10^6/yr		Bone marrow	300	13.6
Medical personnel	Occupational	100,000		Whole body	260–350	0.1
Atmospheric weapons tests	Total population	220×10^6		Whole body	4–5	4–5
Nuclear industry						
Commercial nuclear power plants (effluent releases)	Population within 10 mi	$<10 \times 10^6$		Whole body	≪10	≪1
Commercial nuclear power plants (occupational)	Workers	67,000		Whole body	400[c]	0.1
Industrial radiography (occupational)	Workers	11,250		Whole body	320	0.02
Fuel processing and fabrication (occupational)	Workers	11,250		Whole body	160	0.01
Handling by-product materials (occupational)	Workers	3,500		Whole body	350	0.01

Federal contractors (occupational)	Workers	88,500	Whole body	~250	0.1
Naval nuclear propulsion program (occupational)	Workers	36,000	Whole body	220	0.04
Research activities					
Particle accelerators (occupational)	Workers	10,000	Whole body	Unknown	≤1
X-ray diffraction units (occupational)	Workers	10,000–20,000	Extremeties and whole body	Unknown	≤1
Electron microscopes (occupational)	Workers	4,400	Whole body	50–200	0.003
Neutron generators (occupational)	Workers	1,00–2,000	Whole body	Unknown	≤1
Consumer products					
Building materials	Population in brick and masonry buildings	110×10^6	Whole body	7	3–4
Television receivers	Viewing population	100×10^6	Gonads	0.2–1.5	0.5
Miscellaneous					
Airline travel (cosmic radiation)	Passengers	35×10^{6d}	Whole body	3	0.5
	Crew members and flight attendants	40,000	Whole body	160	0.03
Airline transport of radioactive materials	Passengers	7×10^{6e}	Whole body	~0.3	0.01
	Crew members and flight attendants	40,000	Whole body	~3	<0.001

[a] Annual dose rates from important significant sources of radiation exposure in the United States.

[b] Based on personnel dosimeter readings; because of relatively low energy of medical X rays, actual whole-body doses are probably less.

[c] Average dose rate to the approximately 40,000 workers who received measurable exposures was 600–800 mrems/yr.

[d] Total number of revenue passengers per year is 210×10^6; however, many of these are repeat airline travelers.

[e] About 1 in every 30 airline flights includes the transportation of radioactive materials; assuming 210×10^6 passengers per year (total), approximately 7×10^6 would be on flights carrying radioactive materials.

Table 2. USRPC Estimate of Population Exposure Due to Various Sources of Radiation[a]

Source	Annual collective effective whole body dose—10^6 person-rem/year
Cosmic rays	6
Terrestrial radiation	6
Internally deposited radionuclides	
Radon and progeny (0.004 WL)	10^b
All others	8
Medical diagnostic X rays	10
Fallout	1
Building materials	1
Airline travel	0.1

[a] UNSCEAR—1980 (draft).
[b] Reduction of average air exchange rates in houses by one-half without additional controls would increase this collective dose to approximately 20×10^6 person-rem/year.

ulation doses are integrated over vast expanses of space and time and the linear, non-threshold hypothesis is used to estimate risk. Integration of miniscule individual doses over the population of the world and thousands of years will result in the calculation of substantial numbers of cancers or genetic effects. These numbers are, of course, only upper limits and tell us very little about the actual situations, other than that the actual effect is somewhere between zero and the number calculated. Such integrations do not affect the estimated value of risk to an individual at all. Moreover, they cannot be used to estimate the risks to society without a frame of reference—for example, the total number of people who lived and died in the area of time period of interest. The number of cancer cases calculated by such an exercise is so dwarfed by the number of naturally occurring cancers that it is meaningless.

It should be noted that population dose (collective dose), as a measure of detriment, must be used carefully. For example, is the society worse off when collective dose doubles because of the doubling of the population? The individual (per capita) dose and risk

Table 3. Natural Background Levels of Variability

Area	Population included	Background level (mrem/yr)[a]
U.S. (Atlantic and Gulf Coast)	6,760,000	65–70
U.S. (noncoastal plains)	46,780,000	80–95
U.S. (Colorado plateau)	1,070,000	125–160
U.S. (Leadville, Colorado)	10,000	235
U.S. (central Florida and New England areas)	?	200
Brazil (coastal strips)	30,000	500
France (granite rock areas)	7,000,000	180–350
India (Kerala and Madras States)	100,000	1300
Niue Island (Pacific)	3,000	1000
Egypt (Northern Nile Delta)	Densely populated	300–400
World (calculated average)	2 billion	80–90

Source: NIH Publication No. 80-2087.
[a] Levels given are for external radiation only. Internal doses are not included.

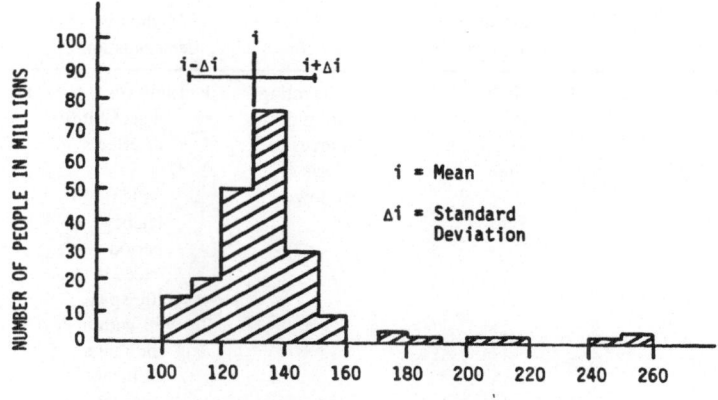

ANNUAL DOSE IN MILLIREMS FROM NATURAL BACKGROUND RADIATION

Figure 6. Variability of state-by-state average natural background levels. (Adapted from Adler and Weinberg, "An Approach to Setting Radiation Standards," *Health Physics 34*, pp. 719–720, 6/78.)

remain the same, and presumably all benefits and available societal resources are apt to double as well.

Another point to remember is that doses of radiation from man-made and man-enhanced sources are always incremental to doses from natural background. Except in extremely unusual settings (e.g., deep undersea, in laboratory shields), the actual dose received by a human being never really approaches zero closer than the minimum level of natural background (about 70 mrems per year).

Summary of Proposals for De Minimis Radiation Levels

In recent years several individuals and organizations active in the radiation protection field have proposed that de minimis levels of radiation dose or related quantities be identified. Table 4 summarizes some of these proposals. They are described in more detail in the Appendix.

Guidance of Expert Bodies

National Council on Radiation Protection and Measurements (NCRP).
NCRP was chartered by Congress in 1964 to, *inter alia*, collect, analyze, develop, and disseminate, in the public interest, information and recommendations about protection against radiation. Council members are recognized experts in the areas of the organization's interests. At the present time, the council's Scientific Committee 1, on Basic Radiation Protection Criteria, is preparing a report on proposed development of the radiation protection system. Among the proposals being considered is the establishment of a de minimis annual dose equivalent limit of concern.[29] At the time of the committee's report at the March 1979 meeting of NCRP, a value for the de minimis level of 10 mrems per year was under consideration.

Scientific Committee 1 is expected to report to the council early in 1982. The council

Table 4. Summary of Previously Proposed De Minimis Levels

Name	Definition and application	Benefit	Method of determination	Suggested quantitative value
Eisenbud (1980)[1]	Level below which exposures are ignored	Avoid devoting effort and resources to analysis of very low doses	1. Limit of detectability of effects 2. Dose at which latency period exceeds life span 3. Well within the range of natural background doses	Whole body external—20– 100 mrems/yr
Rossi (1980, 1981)[24,25]	Level of "no concern"; cutoff for ALARA	Reduction in ambiguity of regulatory interpretation of ALARA practices	Possibly fraction of background	30% of natural background
Whipple (1980)[2]	No observable effects on health	Avoid disproportionate effort in regulating insignificant doses	Comparison with variation in natural risk of disease levels	500 mrems/yr for Low-LET radiation, as low as 100 mrems/yr
Webb and McLean (1977)[26]	Level that individual does not consider in decision making; derived cutoff level for ALARA and optimization	Consistency in practice; allow better allocation of resources to higher dose problems	Based on risk level deemed "negligible" to individual	10 mrems/yr to organ or to whole body; cutoff for a single "practice" at 0.1 mrems/yr
Weinberg and Adler[27,28]	Low-dose radiation standard for individual in population		Standard deviation (weighted with the exposed population) of natural background	20 mrems/yr
NCRP (1979)[29]	De minimis dose. (draft)			10 mrems/yr
ICRP[30,31]	Calculation of collective doses	Concentrate efforts on reduction of doses near the limit	Minimal further contribution to collective dose estimate	Stop calculations when further summations change results by less than a factor of 3

will thereafter consider the recommendations of its committee for adoption as recommendations of the NCRP.

Another committee of NCRP, the Study Group on Acceptable Risk (Nuclear Waste), reported in 1980 that it is developing a report on risk assessment and involvement of the public in such assessments. The draft report includes, as one of the techniques of assessing risks, "attempts to identify and quantify a de minimis level of risk."[32]

International Commission on Radiological Protection (ICRP). At the present time, ICRP, which is the international counterpart of NCRP, has not specifically endorsed a de minimis policy or recommended a de minimis dose. However, the commission apparently accepts some sort of relative de minimis concept on a case-by-case basis, at least insofar as limiting ALARA is concerned.[30] Thus ICRP report 26, in discussing collective dose equivalent, states: "It is often not necessary to assess the contributions from small values of [dose equivalent] accurately, provided that an upper estimate shows that they would not add significantly to the total [collective dose]." A somewhat more quantitative statement of this principle appears in ICRP Publication No. 22, "Implications of Commission Recommendations that Doses be kept as Low as Readily Achievable." Referring to evaluating total detriment from a source of radiation, irrespective of its relation to other detriments such as that from natural background radiation, the report states that "at levels of individual dose that are small fractions of the relevant dose limit, there will be no need to pursue the summation beyond the point where it becomes clear that . . . further contributions . . . will not change the estimate of population dose by more than a factor of about 2 or 3, or . . . beyond the point where it becomes clear that the remaining detriment is insignificant in comparison with the benefits expected from the source. This judgement is made on a case-by-case basis."[31]

ICRP Publication No. 22 also recognizes that "at low levels of individual dose, e.g., those small by comparison with variations in local natural background, the risk to the individual is so small that his health and welfare will not be significantly changed by the presence or absence of the radiation dose." This observation is made, by the authors of ICRP Publication No. 22, not to support a de minimis level, however, but to indicate that in decision making, the concept of collective dose has to be supplemented by consideration of the dose to individuals, and effort has to be devoted more to dose reduction for individual dose levels that are near the relevant limit than to reducing *low levels* further.[31] This represents what we may call a relative de minimis approach.

Possible Scientific Bases for Determining De Minimis Dose Levels

This section discusses some of the methods that have been suggested or might be considered for determining scientific de minimis radiation exposure limits. Ideally, the determination of such a level would be solely a scientific or technical matter. While some technical judgment might be involved in these methodologies, ideally they should involve no consideration of social, political, or economic factors, or determination of acceptability, or public perception. Such considerations are appropriate in the formulation of a regulatory policy and the determination of regulatory cutoff levels. They are discussed in a later section. As a practical matter, however, some of the scientific methodologies depend on assumptions of the acceptability of background related risks to reasonable people.

The methodologies presented in this section would be used by scientific experts in the areas of radiobiology, health physics, epidemiology, statistics, and similar fields. Those described in the section on the legal context would be used by governmental bodies representing the people, and might utilize, to a greater or lesser extent, the results of the expert determination made by the methods described in this section.

It should be noted that the various approaches described in this section are not mutually exclusive. Therefore, in practice, a combination of approaches might be used to determine a value that can be considered de minimis from several points of view.

Methodologies Based Solely on Radiobiological Considerations

Absolute threshold. The most satisfying and the most publicly acceptable type of de minimis dose would be one that is based on an absolute biological threshold, such as is observed, for example, for the acute radiation syndrome. The ideal de minimis level would be one where a strong, well-tested theory, and extensive data on human beings, supported the conclusion that, at that level and below, there was no somatic, teratogenic, or genetic nonbeneficial effect of radiation on any exposed individual or his progeny. As already discussed, at the present time no such absolute de minimis level can be identified, nor is there much prospect of that being done in the foreseeable future. It is also true that the existence or nonexistence of such a level has not been scientifically proven and is unlikely to be proven in the foreseeable future. While the fact that "no absolutely safe level of radiation is known" does not necessarily imply that "no absolute safe level of radiation exists," it precludes, for the present at least, the setting of a de minimis dose that, considered alone, can be unequivocally declared to have been proven to involve absolutely no detrimental effect.

Practical threshold of latency. Radiation doses down to small levels may cause bodily changes that would eventually result in the occurrence of cancer in some exposed individuals. However, if the natural life-span of those individuals is shorter than the latent period before the cancer becomes manifest, those changes are of no concern to the individual or to society. Such small radiation doses would therefore be de minimis. As Dr. Eisenbud discusses in his paper,[1] there are experimental, epidemiological, and theoretical reasons to believe that such a "practical threshold" exists (see also p. 151 and the Appendix). Thus, for low dose rates of low-LET radiation, this concept of a "practical threshold" offers an objective method of de minimis level definition. Dr. Eisenbud recommends that the highest priorities in radiobiological research be given to study of the interrelationships involving dose, dose rate, and latency period, recognizing that there is probably not enough information available now to establish definitive de minimis levels on this basis alone. The information that is available, however, can be used in conjunction with some of the other methodologies described in this section, as input to the determination of de minimis levels. Eventually, if enough information is developed, it may be possible to determine, on this basis, levels of radiation dose rate that would have no adverse effect on the health or well-being of an exposed individual during his lifetime.

Practical threshold of integer effects. As discussed previously, it is possible, using an assumed dose–effect relationship and a population size, to calculate a dose that would result in less than one death in the entire population in some period of time. While the determination of such a value is a scientific task, the use of one death, or any other number, as a criterion, is in the domain of regulatory policy (see p. 192).

Methodologies Based on Comparisons

Detectability. Mankind is daily exposed to radiation,to health effects, and to risks, many of them indistinguishable from those that are or might be produced by regulated sources. Several scientists have suggested that if an absolute de minimis level, based solely on radiobiology, cannot be determined, determination of an incremental level of dose, effect, or risk, caused by regulated sources, can be based on comparison to the "background" of doses, effects, and risks, or to variations or uncertainties in that background. There are two major assumptions involved in most such comparative approaches: (1) that there is no threshold for adverse effects of radiation—i.e., there is a finite risk associated with any dose of radiation, however small or nondetectable it may be—and (2) that the dose, effect, or risk of exposure to background levels, or to levels equivalent to variations or uncertainties in background, are, in some sense, "small" or "negligible."

As discussed, assumption (1) is generally made because its converse, that there is a threshold below which there is no effect, has not been proven. It is generally a "conservative" assumption in that it tends to err on the side of overestimation of radiation effects. Assumption 2 is based on the fact that no adverse effects of living in high background areas have been detected. This assumption brings these methodologies close to the domain of the regulatory decision maker (see The Legal Context), since "smallness" or "negligibility" in this context may imply societal "acceptability."

Nondetectability of Radiation. One might consider a level of radiation dose that is undetectable to be a candidate for de minimis designation. If detection by the human senses is meant, however, the level is certainly too high, and if detection by the most sensitive of instruments (in the absence of background) is meant, the level is too near zero to be of practical interest. A more reasonable approach would be to base the de minimis level on the maximum level of radiation that cannot be detected (with reasonably sensitive instruments) in the presence of background radiation. This detectability "threshold" depends not only on the level of incremental radiation and the accuracy and precision of the instrument used to detect it, but on the pattern of natural and random variations in the background radiation level and the time scale involved. As a statistical concept, this methodology is similar to the method based on comparison to variations in background discussed in a later subsection. A de minimis dose determined on the basis of nondetectability could be characterized as an incremental dose too small to be measured.

Nondetectability of Health Effects. Another approach, related to nondetectability, looks to the threshold of detection of health effects causally related to radiation exposure. There is no way, at this stage of our knowledge, that such effects can be detected clinically in an individual, even at doses far above current dose limits for individuals in the population. Detection might be possible only statistically through epidemiological studies. Thus we express health effects statistically, in terms of risk, as discussed in the section on legal context.

Nondetectability of Risk. Statistical studies are used to determine the risk, or probability of a deleterious effect, due to radiation exposure. Dr. Eisenbud characterizes the de minimis dose level determined through epidemiological studies as one in which the risk is so small that it cannot be measured with today's most sophisticated investigative methods.[1] Of the approaches discussed in his article, he considers this the weakest "on scientific grounds," because "sizeable effects can be buried in the statistical noise with which the epidemiologist must contend." Nonetheless, use of the information on non-detectability, developed by such epidemiological studies, may be helpful in bounding the magnitude of effects and confirming that any effects of a de minimis dose, established on another basis, are not detectable in the population.

Comparison with background

Background Radiation. Comparison with background radiation has often been used to place other sources of radiation in perspective.[1] Dr. Eisenbud discusses three measures of background radiation exposure that can be useful for such comparisons: whole-body dose, which is usually also a measure of gonadal dose, from external radiation and radiation from within the body; the dose to bone from radium-226, used for comparison with bone-seeking radionuclides like strontium-90; and the dose to the respiratory tact from radon and its daughters, used where there is exposure to airborne radionuclides.

Dr. Eisenbud proposes that radiation doses of the order of geographical variations in natural background should be considered to be de minimis (see Table 3). He reasons that people do not take background exposures into account when deciding where to relocate; therefore, they must consider any dose increments involved to be *negligible*. Such an *acceptability* rationale (see The Legal Context), raises some problems, however. Most people are totally unaware of the existence of background radiation, let alone its geographical variations and possible correlation with health risks. Thus, it is not precise to say that they consider these exposure differences trivial; their neglect to consider them may arise totally from ignorance. Of course, if there were clearly detectable and widely know health effects associated with these changes, any subsequently observed acceptance behavior would support the de minimis designation. Here, however, there is no detectable effect and, as Dr. Eisenbud noted in another context, nondetectability, even by sophisticated epidemiological studies, is not a guarantee that there is no deleterious effect.

The acceptability argument would be a stronger one if it related to acceptability of geographical background variations by relocating and vacationing radiobiologists, radiologists, and health physicists, experts who may be presumed to be knowledgeable both about the existence of variations in background radiation and about the potential associated risk levels, rather than by the general public. A sociological study of such behavior might buttress the position that this is a *scientific,* not a *political,* approach. The rationale would then be one of expert opinion of acceptability of risk as revealed through the behavior of the experts in their private lives.

Another rationale for the use of background level variations as de minimis might be developed from consideration of man's evolution. Man has evolved over the eons in a background of spatially and temporally varying radiation levels. The species is apparently adapted for survival in this environment. Whatever genetic effects have resulted, they have not adversely affected species' survival or development. If levels of radiation of the order of variations in background threatened his existence, man would have evolved senses for detecting radiation and avoiding dangerous levels, or the species would have died out. While such an argument may be useful in showing that natural background radiation levels are not sufficiently life-threatening to affect species' survival, it may not necessarily imply that those levels carry so little risk as to be negligible to an individual.

In its report, the American Physical Society (APS) Study Group on Nuclear Fuel Cycles and Waste Management stated it considered natural background radiation as a reasonable basis for comparison. Note that fluctuations in background exist in time and in space that may be comparable in magnitude with background itself, the report's authors reason that "since, in the course of human existence no statistically discernable effects have been associated with such fluctuations, it follows that the effects of increments in dose and dose rate small compared to background fluctuations will be small compared to an already undetectable level of effects."[33] The APS study also cites the earlier BEIR committee (1972) recommendation that natural background radiation be used as a standard for comparison for genetic effects of radiation.[34]

Once it is determined that background is to be used as the basis for comparison, it must then be decided with what aspect of background, and that measure of its variability or uncertainty, a de minimis dose is to be compared. For example, Adler and Weinberg (see Appendix) have used, as a measure of background level variability, the standard deviation of the distribution of average background radiation dose, state by state, in the United States, to determine a dose that is "small compared to background" (see Figure 6).

Among the possible distributions of background, the variability of which might be used, separately or in combination, in determining a de minimis dose are:

- Distribution of state average background levels over the United States.
- Distribution of background levels over the U.S. population.
- Distribution of spatial average U.S. background levels over some reasonable length of time.
- Distribution of background levels at a typical site over time.

Uncertainties in measurements and estimates of background doses might also be used to determine a de minimis level. Such uncertainties include instrumentation limitations and uncertainties in estimates of shielding by the body, building, and vehicles; proportion of time spent indoors and outdoors by individuals; and the effect of other lifestyle variations.

Another consideration in the use of background for comparison is its definition; should medical diagnostic and other man-related exposures be included in *background* along with its *natural* component? Dr. Eisenbud has stated his opinion that medical exposures should not be included, because there is a different cost-benefit balance involved.[32] Others believe, however, that for small diagnostic and research doses, no risk-benefit analysis is actually done; the risk is assumed to be negligible,[26] and therefore a de minimis approach might be appropriate for such doses.

The *evolutionary* argument might suggest that only *natural* radiation be considered. The *detectability* rationale, on the other hand, would demand that all competing background risks be included, and would theoretically require all current sources of background exposure to be considered. A problem the latter rationale might raise is one of positive feedback; it exacerbates the problem of cumulative de minimis doses (see Development of a Regulatory Cutoff Policy). The higher the total background is, the greater the level of comparative de minimis dose is likely to be. If de minimis doses accumulate, eventually the background level and its variability might be further increased and the de minimis level along with it. Theoretically, at some point, if enough additions accumulate, the total background could reach a level that is no longer *acceptable*. While this eventuality may not be a practical likelihood, the argument is logically irrefutable. The solution is probably, as discussed later, to require the regulatory authorities to monitor total long-term background increases, and reevaluate the reasonableness of a de minimis policy periodically.

Background Health Effects. In human society, illness, disability, and death are caused by a variety of natural and man-made agents and are, indeed, inherent in the process of life itself. One approach to establishing a de minimis dose would be to select a dose level that results in a total number of deleterious health effects in the population that is *small* compared to the number of naturally occurring effects and those caused by other environmental agents. Comparisons based on such considerations are usually expressed in terms of risk (probability of deleterious effects) rather than number of effects alone (but see p. 162). The use of background health risk as a basis for comparative determination of de minimis dose is discussed in the next subsection.

Background Risks. Man lives in a world of risk. One approach to determining a de minimis dose is to consider that, by definition, the quantitative risk of averse health effects, believed posed by a de minimis dose, is negligible, as far as the individual is concerned, compared to the other risks he faces. Risk statistics can then be used to determine the level of risk that would be negligible. In this section we consider objective determinations of the negligibility of risks. The subjective concept of *acceptable* risk is discussed in the next section.

If an incremental risk cannot be detected in the statistical noise of the background risk of illness or death to which the individual is subjected, it can be considered de minimis. The rationale is that the individual cannot take a risk into account in decision making if he cannot detect it. This approach is analogous to that described above for comparison to background radiation and is subject to the possible criticisms stated previously that substantial effects might be masked by such noise.

One can use this comparative approach with a variety of risk data: risk of illness or death, risk of cancer, risk of possibly radiogenic types of cancer, risk of genetic effects. As with background radiation, various distributions over space and time can be used as a basis for determining measures of variability. Dr. G. Hoyt Whipple has evaluated a de minimis dose level based on the variability over time of total annual U.S. cancer fatality risk statistics.[2] Once a level of negligible risk has been determined in this way, it can be correlated, using a dose–effect relationship of some type, with a radiation dose. That dose is then the de minimis level (see Figure 7).

There are a number of ways in which the level of negligible risk can be determined by this method. As noted above, various types of risks might be used for comparison. Variability over space, time, and life-style could be considered, as well as uncertainties such as those related, for example, to diagnosis or reporting.

Relative contribution to total regulated incremental dose. This methodology considers doses in comparison only to other regulated doses or risks, without regard for unregulated competing background doses or risks. An example of this approach is ICRP's suggestion that in calculating collective doses, doses so small that their inclusion

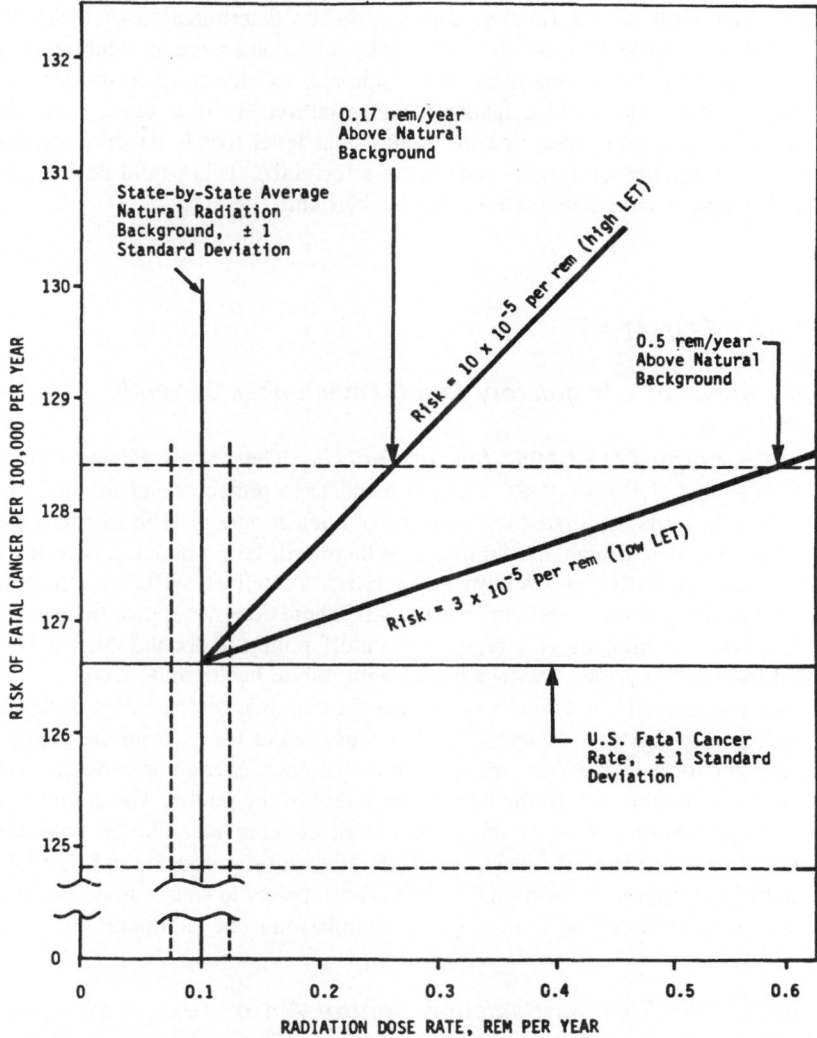

Figure 7. Calculation of de minimis dose based on comparison with background cancer risk. (Adapted from G. H. Whipple, "A Practical Threshold for Radiation.")

would not change the result by more than a factor of 2 or 3 need not be included in the calculation.[31] This is clearly a case-by-case determination that serves to cut off the only low-dose *tail* on a distribution of doses over a population. The ICRP guidelines appear to require the consideration of all doses, however small, if their distribution over a population is relatively uniform.

Another relative approach would consider calculational uncertainties in estimated dose calculations. There are uncertainties associated with input parameters, modeling assumptions, and numerical calculation methods. These uncertainties could be used as a basis for estimating a level of dose that is negligible by comparison.

Another example of the relative approach is the determination of levels that are negligible with respect to other levels produced by the same source or radiation utilization program. The qualitative aspects of such an approach are described in the section on p. 189. Quantitatively, one could consider the distribution of doses due to, say, nuclear power plant effluents and determine an incremental level that is de minimis based on variations in that background over space or time (or both). This would be a level that is negligible because it is undetectable in that background.

THE LEGAL CONTEXT

Appropriateness of a Regulatory Cutoff Level under Currently Applicable Law

Atomic Energy Act of 1954 (as Amended).   The Atomic Energy Act of 1954 (the Act), as amended,[35] gives NRC a broad mandate to protect the health and safety of the public from hazards associated with the use of nuclear energy. The standard has been interpreted as one of reasonable assurance that there will be no undue risk to the public from NRC licensed activities. Within these broad guidelines NRC has discretion in establishing regulatory policy and implementing it. There does not appear to be any reason why NRC could not implement a regulatory cutoff policy under the current law, if it determined that such a policy was consistent with public health and safety.

The Act (Section 81) specifically authorizes the commission to exempt certain classes or quantities of "by-product" *material,* Or kinds of uses or users, from the requirements for a license set forth in the Act, when it finds that such exemption would, *inter alia,* not pose an unreasonable risk to the health and safety of the public. The commission has done this by regulation. Most of the materials of concern in radiation protection are defined as by-product material (fission products, activation products, and mill tailings).

A similar exemption, in Section 62 of the Act, applies to source materials (uranium and thorium) in quantities determined by the commission to be "unimportant".

Uranium Mill Tailings Radiation Control Act of 1978 (UMTRCA).   The Uranium Mill Tailings law[36] has two major sections. The first sets a remedial action program for existing unlicensed uranium mill sites, while the second brings mill tailings under NRC licensing authority by expanding the previous definition of by-product material in the Atomic Energy Act to include those materials.

In the Mill Tailings Act, Congress finds that mill operations "may pose a potential and significant radiation health hazard to the public," and that tailings should be controlled "to present or minimize radon diffusion into the environment" [Section 2(a)]. A purpose of the two programs of the Mill Tailings Act is "to minimize or eliminate radiation health hazards to the public" [Section 2(b)].

The remedial action part of the Mill Tailings Act is to be carried out by the Energy Department. However, the secretary of energy must rely primarily on the advice of the administrator of the Environmental Protection Agency (EPA) in assessing the relative potential health hazards of the sites of concern [Section 102(b)]. The law also authorizes the administrator of EPA [Section 206(a), adding Section 275 to the Atomic Energy Act] to issue standards for the protection of public health, safety, and the environment from radiological and nonradiological hazards associated with tailings. Under Section 205(a) (adding Section 84 to the Atomic Energy Act), NRC is given authority to ensure that the management of tailings is carried out in a manner it deems "appropriate to protect the public health and safety and the environment from the associated radiological and non-radiological hazards."

In addition to bringing tailings under NRC licensing jurisdiction, the law permits an Agreement State to take over regulation of mills and tailings and set standards, for protection of public health, safety, and the environment from the hazards associated with the material, which are equivalent to, or more stringent than, standards adopted by NRC and the EPA.

A regulatory cutoff policy for exposure to radiation from mill tailings, applied by EPA and NRC, would probably be consistent with the Mill Tailings Act, as long as it could be characterized as *minimizing* the associated hazard.

National Environmental Policy Act (NEPA). The National Environmental Policy Act (NEPA) requires that for every major federal action significantly affecting the quality of the human environment,[37] a detailed impact statement be prepared. The federal agency preparing the statement must consider environmental impacts, unavoidable effects, alternatives to the proposed action, and other aspects of the action.

The NRC has developed regulations covering its NEPA activities. These require a balancing of costs and benefits and quantitative evaluation of alternative actions where possible. It should be noted that the agency must make threshold determinations of whether actions are major or have a significant impact. The statutory language and court interpretation of this statute support a regulatory cutoff for negligible or insignificant effects.

Current and Proposed Regulations Involving Regulatory Cutoff

Nuclear Regulatory Commission. Table 5 summarizes some of the current regulatory cutoff levels applicable to various aspects of NRC regulation.

Other Federal Agencies. Table 6 summarizes some of the current and officially proposed regulatory cutoff levels applicable to regulation by the Department of Energy (DOE) and the Environmental Protection Agency (EPA).

Current Proposals for Regulatory Cutoff Levels

This section describes some of the current proposals for cutoff levels.

Nuclear Regulatory Commission Views. The following sections summarize the views of various components of the NRC as expressed in recent publications and proceedings.

NRC commissioners. In a memorandum opinion issued in March, 1981, the commission (Commissioners Hendrie, Gilinsky, Bradford, and Ahearne) denied a petition,

Table 5. Current Regulatory Cut-Off Levels—NRC

Section of Title 10 Code of Federal Regulations	Description of regulation
PART 19 Notices, etc., for workers	
19.13	Notification applies only if data need to be recorded (see 20.202 below)
PART 20 Standards for protection against radiation	
20.202	Personnel monitoring needed only if dose is likely to exceed 25% of MPD for calendar quarter (1250 mrems) for worker or 5% of MPD for minor
20.203(d)	No posting needed if less than 10 times Appendix C quantities in area or room
20.203(f)	No posting needed for containers with less than Appendix C quantities
20.204	No posting needed if sealed source emits less than 5 mrems/hr at 12 inches
20.303	Disposal into sanitary sewers is permitted if quantity is below Appendix C, below concentrations in Appendix B, or below one Curie per year
20.401	Exposure records needed only for personnel in restricted area (see 20.202 above)
20.407	Personnel monitoring reports needed only for personnel in restricted area (see 20.202 above)
PART 30 By-product material licensing	
30.14	Persons exempt from licensing if concentrations are less than 30.70, for Parts 31–35 of 10 CFR
30.15, 30.16 30.18, 30.19 30.20	Stipulate exemptions of specific devices, resins, quantities per Schedule B of 30.17, self-luminous products, gas and aerosol detectors
30.70	Schedule A lists exempt concentrations
30.71	Schedule B has a listing of quantities that are exempt
PART 32 Manufacturing licenses	
32.24	A table of organ doses sets forth safety criteria that must be met by manufactured products, under 32.23 conditions depending upon the possible failure rate of the products
32.28	A table of organ doses sets forth safety criteria that must be met by manufactured products, under 32.27 conditions depending upon the possible failure rate of the products; Subpart B of Part 32 includes a long list of conditions that permit a manufactured product to be transferred to persons generally licensed; 32.51 conditions are based on the use of persons "not having training in radiological protection," and, under ordinary conditions, it is unlikely that any person will receive a dose in excess of 10% of MPD in a calendar quarter (1250 mrems) for workers or a dose commitment greater than Column IV in 32.24 under accident conditions
PART 40 Source material	
40.13	Specifies "unimportant" quantities

Proposed regulations

PART 51 Environmental protection—Licensing and regulatory policy (March 4, 1981)	
51.20	No further discussion of the environmental effects addressed in Table S-3 is required in applicant's environmental report
51.23	The commission has found that fuel cycle impacts addressed in Table S-3 cannot significantly affect the cost–benefit balance for a light water reactor; except for radon-226 and technetium there shall be no further considerations of the impacts addressed by Table S-3
PART 61 Disposal of low-level waste (July 24, 1981)	
	(see text p. 173)

Table 6. Current Regulatory Cutoff Levels—DOE and EPA

Agency	Regulation section	Description
Department of Energy	10 CFR712	Grand Junction remedial action criteria
	712.6	Criteria for remedial action levels of external gamma radiation (0.05 mr/hr) and radon daughter concentration (0.01 WL) below which no remedial action is indicated
Environmental Protection Agency	40CFR190	Environmental radiation protection standards for nuclear power operations
	190.10(a)	Limit of 25 mrems/yr whole body, 75 mrems/yr to thyroid, 25 mrems/yr to any other organs of any member of the public, as a result of planned discharges from uranium fuel cycle facilities to the environment
	190.10(b)	Limit on quantity of radioactive materials entering the environment for uranium fuel cycle activities per gigawatt year: 50,000 Ci Kr-88, 5 mCi I-129, 0.5 mCi Pu-239 and other alpha emitting transuranic elements with half-life greater than 1 year
	40CFR192 Part B	Uranium mill tailings cleanup of contamination from inactive sites
	192.12(a)	No remedial action needed if average contribution of Ra-226 in any 5-cm thickness of soil within 1 foot of surface, etc., is less than 5 pCi/gm
	192.12(b)	No remedial action if level in building is less than 0.015 WL (radon decay products)—including background, and 0.02 mr/hr (indoor gamma radiation)
	192.12(c)	No remedial action if cumulative lifetime dose to any organ of maximally exposed individual is less than the maximum dose equivalent due to Ra-226 and its decay products under 191.12(a) and (b)

filed by operators of uranium mills and the state of New Mexico, to stay the commission's uranium mill licensing requirements.[38] One of the contentions of the mill operators was that the NRC has admitted that the risks posed by uranium milling are de minimis, citing NUREG 757, the so-called radon report.[39] This contention, the commissioners asserted in the memorandum, is "wrong."[38]

To support their assertion, the commissioners stated: "Radon is the primary source of long-term public exposure to radiation resulting from uranium milling. If adequate measures are not taken to control radon emissions from mill tailings piles, the public exposure to that source would exceed its exposure to all the other radiation sources associated with the uranium fuel cycle. The radon report states that radiation exposure due to untreated piles will be one hundred times the exposure from piles stabilized in accordance with NRC regulations." Given the logic of this rationale, it appears that the commission is using a purely relative standard for determining whether something is de minimis. That is, a risk cannot be de minimis unless it is small in relation to the risk posed by other radiation sources in the uranium fuel cycle, regardless of what actual level of risk the other sources pose and their magnitude compared to background. From the perspective of the commissioners, as expressed in this memorandum, something larger than a regulated level (or a level proposed for regulation) cannot be de minimis by definition.

The maximum radon emanation rate set by NRC for mill tailings disposal areas is 2 pCi per m²-sec. This is within the range of natural radon emanation rates from soil, 0.65 to 2.3 pCi per m²-sec.[40] The NRC staff concluded that, at its limiting value, the dose from radon emanating from a tailings pile would exceed background by about 14 mrems per year at a fence post 100 meters downwind from the tailings.

The annual dose from outdoor radon in the western United States is reported to range from 0.37 to 1250 mrems, with an average of 150 mrems. Typical levels inside houses range from 150 mrems per year (trailer home in western United States) to 1100 mrems per year (mud-lined Indian huts in western United States).[41]

NRC licensing and appeal boards

The Perkins Case and Other Radon Cases. In the licensing proceedings for Duke Power Company's Perkins plant, the presiding Atomic Safety and Licensing Board considered the environmental impact of radon released to the environment due to the mining and milling of the fuel for the plant, in the context of its significance in a NEPA cost-benefit analysis.[42] Discussing the calculated effects of the radon released, the board said: "A possible half a death per year in a population of 300 million people is a minimal impact. Under NRC stabilization procedures and reasonable regulation on open pit reclamation, the impact will be 100 times less." This certainly appears to be a statement that a particular risk is absolutely de minimis. The board goes on to apply a comparative standard as well, stating, "we find that the best mechanism available to characterize the significance of the radon releases associated with the mining and milling of the nuclear fuel for the Perkins facility is to compare such releases with those associated with natural background. The increase in background associated with Perkins is so small compared with background, and so small in comparison with the fluctuations in background, as to be completely undetectable. Under such circumstances the impact cannot be significant."

In a discussion of the Perkins record, the consolidated appeal boards considering the same issue in several other cases, in their majority opinion[43] characterized the licensing board's decision in Perkins as a "de minimis approach for assessing the eventual significance of possible health effects from [radon] releases." They declared that before the validity of that approach could be considered, intervenors had to be given an opportunity "to challenge the factual underpinnings of the Licensing Board's de minimis rationale." Therefore they called for further proceedings on the "health effects" issues.

In a dissent, appeal board members Drs. Buck and Johnson stated, "in circumstances such as this, in which the addition of a natural environmental substance (i.e., radon) caused by human activities is extremely small compared with the existing natural concentration (it is small even compared to fluctuations in that concentration), we believe that any assignment of environmental impact of the incremental addition could only be characterized as remote and speculative. We conclude that this impact may properly be ignored in the assessment of the overall environmental impact of a nuclear power plant."[43] Noting that the appeal board, in an earlier opinion in the same cases and said, "if we were to subscribe to [the Perkins Board] view, there would appear to be no reason to consider the question of health effects further," the dissenters expressed the opinion that no further evidence would be needed to decide the de minimis question.[44] This case is still pending before the appeal boards as of this writing.

The Maine Yankee Decision. The Appeal Board in the Maine Yankee case[45] sanctioned the use of background radiation as a measure of "the dimensions of [radiation] exposure in normal operations." The board stated: "All of the FES [final environmental statement] discussion of radiation effects of normal plant operations is in terms of the dosages which will be received by individuals at particular locations, or by the general population. These dosages are then compared to dosages normally received by the same individuals or the general population from natural background. They concluded, insofar as radiation exposure is concerned, it was enough that the FES made clear the dimensions of that exposure . . . and conveyed the message that there would be no significant environmental impact resulting therefrom" (ALAB-161). In the Maine Yankee FES, the AEC staff had compared the conservatively estimated 17 person-rems per year to the population within 50 miles of the plant site due to plant operations, with the 73,000 person-rems per year realistically estimated dose to the same population from natural background.

The Board opinion in the Maine Yankee case does not, however, represent a clearcut adoption of a comparison to background as the standard for determining that a dose is negligible. That board also agreed that a stipulation, which the board characterized as reflecting that "while, as the FES demonstrated, the possibility of a meaningful environmental impact from [such] exposure was very remote . . . that possibility cannot be ruled out entirely," should be added to the FES for the purpose of "full disclosure." They concluded, however, that the denial of an operating license would not be "warranted by the fact that, even though remote, the possibility of meaningful environmental impact from radiation exposure cannot be entirely dismissed." This seems to imply that a dose, small compared to background, might still represent a *significant* risk. It appears that the board is agreeing that if the absolute risk of radiation dosage, from natural and man-made sources, is found to be substantially greater than heretofore believed, the relative standard of negligibility would be superseded by an absolute one.

Thus, the board seems to be using a comparative standard for the level of risk, but considering the possible application of an absolute risk standard as well.

NRC staff

Low-Level Waste Disposal. In its announcements of proposed rules for low-level waste disposal (10CFR Part 61) in February 1980, the NRC staff stated that it is "particularly interested in establishing a de minimis level (a level of radioactivity in waste that is sufficiently low that the waste can be disposed of as ordinary, non-radioactive trash) for short-lived radioisotopes commonly used in medical research and other applications."[46]

Thereafter, the former Radiation Policy Council (RPC), in a September 1980 policy paper, stated that there are waste steams with such small risks that control for radiation protection purposes is not necessary. RPC asked NRC to develop a plan for identification of such streams, starting with carbon-14 and tritium in medical wastes.[47]

In March 1981, NRC staff issued regulations that permitted the disposal of tracer levels of radioactive tritium or carbon-14 without regard to radioactivity. NRC estimated that this change in regulations would result in a maximum dose to an exposed individual likely to be less than 1 mrem per year (based on conservative assumptions), resulting in an estimate of 0.4 health effects in next 1000 generations.[48,49]

NRC staff stated that the limit of 0.05 mCi/g was based on a consultant's recommendation of a level that would cover most research involving tracer use in animals. It also more than covers most scintillation fluid applications. The amount of radioactive material that can be released to uncontrolled sewers was also increased, on the basis that the increased health risk "appears low" (citing EPA comments) and there will be some savings in administrative costs to licensees and some savings in burial site space. The Staff expressed the position that these are benefits to public that should not be foregone.[48]

Disposal or On-Site Storage of Thorium or Uranium Wastes from Past Operations. Effective January 28, 1981, NRC regulations in 10 CFR 20, "Standards for Protection Against Radiation," were amended to delete Section 20.304, which provided general authority for disposal of radioactive materials by burial in soil.[50] Under the amended regulations, licensees must obtain specific NRC approval for such disposal of radioactive materials. In the Federal Register Notice, NRC stated its belief that case-by-case reviews were needed to assure that burial of radioactive wastes would not present an unreasonable health hazard at some future date. The deleted provision of Section 20.304 permitted burial of up to 100 millicuries of thorium or natural uranium at any one time, with a yearly limitation of 12 burials for each type of material at each site. The disposal standards specified that burial be at a minimum depth of four feet, and that successive burials be separated by at least six feet. Thus, a total of 1.2 Curies of these materials could formerly be disposed of each year by burial. The implication is that the NRC staff no longer considers these amounts to be de minimis, although an explicit radiological basis for the change was not presented.

Under the amended regulations, an applicant who wants to bury radioactive wastes must demonstrate that local land burial is preferable to other disposal alternatives.

NRC recently published a Branch Technical Position, setting forth the guidelines for approval of applications for disposal or storage of natural thorium and its daughters, depleted or enriched uranium, and uranium ore and daughters.[51] Concentrations for uncontrolled burial were set on the basis of EPA guidance for transuranics and for uranium processing sites (1 mrem per year to lung, 3 mrems per year to bone from inhalation and ingestion under any foreseeable use of property) and proposed cleanup standards for inactive uranium processing plants (10 mrems per hour above background external dose rate). For natural uranium ores having daughters in equilibrium, the concentration limit is equal to that set by the EPA for radium-226 and its decay products (i.e., 5 pCi/gm, including background). It should be noted, however, that background soil radium-226 concentrations range from 0.1 to 4 pCi/gm. Florida sands contain up to 9 pCi/gm.[52]

The levels for the various materials that are considered "acceptably low" for unrestricted disposal by burial are as follows: natural thorium, 10 pCi/gm; depleted uranium, 35 pCi/gm; enriched uranium, 30 pCi/gm; natural uranium ores, 10 pCi/gm.

In light of their definition as levels that are "acceptably low" for burial with "no restrictions," these might be considered regulatory cutoff levels of the type defined in this report. It appears, however, that these levels are not considered *negligible* insofar as ALARA is concerned. The notice states that "it is expected that any soil contamination will be reduced to levels: as low as reasonably achievable." This seems to imply that in certain cases the published "acceptably low" levels might not be acceptable.

Decommissioning. NRC Regulatory Guide 1.86,[70] on termination of licenses, includes a table of acceptable surface contamination levels for various nuclides. Guidelines published November 1976 include a similar table with a footnote that states: "The average and maximum radiation levels associated with surface contamination resulting from beta-gamma emitters should not exceed 0.2 mrad per hour at 1 cm and 1.0 mrad hour at 1 cm, respectively, . . ." The basis for establishing the values presented is not given in the document.

In 1978 NRC published its "Plan for Re-evaluation of NRC Policy on Decommissioning Nuclear Facilities," and later that same year, a modified plan based on comments received.[68]

A draft report, NUREG-0613, published by the NRC's Office of Standards Development in 1979, discusses residual radioactivity levels for decommissioning. This draft report by Mr. E. Conti includes a disclaimer that "any opinions are those of the author and do not represent official NRC Policy. The report notes the lack of any authoritative definition of a de minimis dose (i.e., a dose equivalent corresponding to a risk that is comparable to the risk from other activities that are generally accepted without special concern)."

The NRC technical staff approach to developing residual activity limits, described by Mr. Conti, is based on the following criteria: (1) Residual activity limits should present a small radiation risk, (2) the limits should allow an effective measurement program to demonstrate compliance, and (3) the limits should be consistent with existing guidance. The staff analysis discussed in the draft report indicates that residual activity levels which would be expected to result in doses of 5 mrems per year to an individual, under realistic exposure pathway conditions, would be both consistent with existing guidance and capable of being monitored for enforcement.

The NRC staff report also includes table of detection limits for direct survey with portable instruments. The gamma detection limit is given as 2.5 to 5 microrems per hour, depending on background radiation levels that typically range from 5 to 10 microrem per hour. This detection limit represents a 50% increase over background (a dose rate of 5 microrems per hour is equivalent to an annual dose of about 44 mrems).

Updating of Radiation Regulations, 10CFR20. The NRC has announced that it plans to revise its regulations on radiation exposure (10CFR Part 20). Although a draft has yet to be publicly issued, it is understood that the concept of de minimis dose and regulatory cutoff are included in current agency staff draft.[53]

With regard to individual doses to members of the public, there would be two practical cutoff levels below the 500 mrems per year individual dose limit. The higher cutoff, at 100 mrems per year, would define a region (100–500 mrems per year) within which a licensee must provide an ALARA program for off-site exposure. The low cutoff, at 25 mrems per year, would define a level below which no reporting of doses to NRC (or any other regulatory action) would be required.

With regard to evaluation of collective doses over time and space, for activities that release radioactive materials to the environment, the staff draft is believed to propose a de minimis level of 0.1 mrem. Using the usual conservative dose–effect estimates, this level represents a risk of radiation-induced cancer of about 1 in 100 million. Such a dose received yearly would result in a risk of less than 1 in 1 million over a lifetime.

Views of Other Persons and Organizations. The following sections describe proposed de minimis dose or risk levels that are based on factors beyond the solely scientific ones presented in The Technical Context and in the Appendix. They provide examples of possible approaches to a regulatory cutoff policy.

Atomic Industrial Forum (AIF). A study was carried out for AIF in 1978 to determine a value for allowable activity levels in solid wastes that would not require regulatory control or surveillance.[54]

The report defined "de minimus (*sic*) levels as concentrations or quantities of potential pollutants that are present in so minute amounts . . . as to be relatively harmless to man and his environment. A waste . . . that meets the criteria defining de minimus levels should be acceptable for disposal or release without any special precautions or monitoring programs."

This study took existing radiation standards and guidelines as *given* and developed the *de minimus* regulatory cutoff levels based on them. In discussing the results of the study, the report stated that the 1-mrem-per-year level derived was selected on a *very conservative* basis and higher levels would likely be appropriate for some applications.

Consideration was given to whether applications of the ALARA principle would result in limits lower than those resulting from the calculations. The study concluded that it did not. The determination of the cutoff level was based on NRC regulations 10CFR50, Appendix I, and 10CFR20.

The study report noted that the dose rate to the maximally exposed person was the basis for the cutoff level derived. It stated: "At such low individual dose rates, a cost-benefit interpretation based on an estimate of the population integrated dose equivalent (collective dose) would be of questionable value in providing guidance."

According to the AIF report, the range and variation in exposures to natural sources of radiation experienced by the population of the United States suggested that a cutoff dose rate of a few mrems per year to individual members of the public would be appropriate. "For example, 3 mrems per year is well within the variation among residents of a given area and is only a few percent of the variation among large groups of people in different parts of the U.S."

The study authors reasoned that "since wastes with activities below the cutoff would not be subject to environmental surveillance after disposal, lower activity limits in such wastes than in monitored reactor effluents (10CFR80 Appendix I) may be justified logically," even though, in view of variations in natural background levels, Appendix I design guides may be too low in the first place (see p. 183).

The AIF report also referred to EPA regulations, 40CRF190. These regulations set an annual limit of 25 mrems to whole body to population members from uranium fuel cycle activities. These regulations exclude operations at waste sites, but do not specifically exclude waste itself. Another EPA recommendation referred to is the Proposed Federal Guidance on limits applicable to transuranium elements, which sets a limiting dose rate for alpha-emitting transuranics of 20 mrems per year.

The AIF report summarizes the criteria it considered for determining a cutoff dose rate level. They are as follows:

1. Level below which the dose rate to maximally exposed individuals is insignificant compared to that typically experienced by a member of the population from natural background ($+/-$ 3 mrems per year, for instance, is not considered significant).

2. Level compatible with existing standards and regulations. Therefore, it must be less than Appendix I 10CFR50 design guides (about 5 mrems per year).
3. Guidance deduced from 10CFR20, Appendix B, which states that radioisotopes in mixtures may be ignored if present in less than one-tenth of its maximum permissible concentration (mpc), up to 25% for a mixture.
4. In the AIF study, the approximately 3 to 5 mrems year dose rate of Appendix I was combined with the 25% of Appendix B, resulting in cutoff levels of 1 mrem per year whole-body dose and 3 mrems per year to any organ.

Dr. Cyril Comar.   The late Dr. Comar of the Electric Power Research Institute (EPRI) presented his views on "A Pragmatic De Minimis Approach" to risk in an editorial in *Science*;[55] it is reprinted in this volume with the permission of the AAAS. He suggested that, to deal with risks on an objective basis, "we . . . define actuarily the existing state of well-being and calculate effects on it." The editorial then sets forth the following statistics for the U.S. population: "In 1975, 1.89 million died out of a population of 213 million, giving an overall probability of 1 in 113. For some specific age groups the values were: 1 to 4 years, 1 in 1425; 5 to 14 years, 1 in 2849; 25 to 34 years, 1 in 692; 55 to 64 years, 1 in 67."

Dr. Comar then discusses the question "What does adding a small risk do to a person's existing probability of dying?" Given the statistics presented, he calculates that, if a young child were exposed to an additional risk of 1 in 100,000 (0.014 in 1425) in 1975, his overall risk for that year would be 1 in 1425 plus 0.014.

For the purpose of discussion, Dr. Comar presented some guidelines in terms of numerical risk, as follows:

1. "Eliminate any risk that carries no benefit or is easily avoided." (This criterion may not be necessary for a regulatory cutoff level as defined in this chapter. A level may be set so low that whether or not its elimination causes a minute benefit would be of no regulatory concern; see Eisenbud in the Appendix).
2. "Eliminate any large risk (about 1 in 10,000 per year or greater) that does not carry clearly overriding benefits."
3. "Ignore for the time being any small risk (about 1 in 100,000 per year or less) that does not fall into category 1." (This is Dr. Comar's quantification of negligible risk.)
4. "Actively study risks falling between these limits (one in 10,000 to one in 100,000), with the view that the risk of taking any proposed action should be weighted against the risk of not taking that action."

The editorial notes that the suggested guidelines are a gross oversimplification, and Dr. Comar describes some of the problems inherent in the adoption of such a policy. "The unfortunate, overtaken by a one-on-a-million catastrophe, have a 100% chance of harm. The hard fact is that attempts to eliminate risks for the unfortunate few tend to markedly increase them for the rest of a large population. This idea is most difficult to defend politically, especially when the unfortunate few are known and the fortunate many are nameless. In addition, it is necessary to take into account such matters as validity and uncertainty in risk estimates, nonlethal and esthetic effects, voluntary versus involuntary risks, societal abhorrences, and the strange versus the familiar."

Nevertheless, the editorial concludes that such a pragmatic de minimis approach would serve to "promote understanding about how to deal with risk in the real world; encourage identifiers of risk to provide risk estimates; focus attention on actions that can effectively improve health and welfare and at the same time avoid squandering resources in attempts to reduce small risks while leaving larger ones unattended; and prevent anxiety, apathy, or derision as a response to the increasing recognition that we apparently live in a sea of carcinogens."

Using a conservative estimate of 10^{-4} as the risk of fatal cancer per person-rem, Dr. Comar's guidelines of 10^{-5} as the level of small risk that can be ignored would imply a regulatory cutoff dose level of about 100 mrems per year.

ORNL decommissioning monitoring study. In 1981, C. F. Holoway et al., of Oak Ridge National Laboratory,[52] issued a report addressing the final steps needed to ensure that a site that has been decontaminated can be released for unrestricted use. The report discusses the monitoring needed to determine a background level at a particular site so that future measurements that are above that level will have a small likelihood of being background measurement and a large likelihood of reflecting contamination. It discusses the variability of background measurements from point to point and time to time, concluding: "Since external radiation exposures in the U.S. range from about 50 to 150 mrems per year (terrestrial and cosmic) a variation of perhaps 25 mrems per year (one-fourth of natural background) would suggest a verifiable figure of 25 mrems per year, increasingly difficult to justify in progressing from 25 to 10 to 5 mrems per year when trying to establish a realistic cleanup value for soil."

Potential Applicability of Regulatory Cutoff Levels in NRC Regulatory Activities

This section considers possible applications of de minimis and regulatory cutoff doses for various aspects of NRC regulatory functions. It should be noted that while the basic regulatory cutoff level would refer to individual exposures it may be necessary to set the cutoff levels in particular applications at different values, for example, to take account of possible cumulation of doses from several sources.

Maximum Permissible Doses and Other Dose Limits. The NRC is currently engaged in revision of its regulations on protection from radiation, 10CFR Part 20 (see p. 174). The current regulations set forth maximum permissible doses for individuals in the public and for those occupationally exposed. Limits for monitoring, disposal, etc., are also specified.

A revised 10CFR Part 20 could specifically adopt a de minimis cutoff policy and set forth an additional dose level as the regulatory cutoff level which would apply generally, except for specifically stated different cutoff values for a particular situation. One or more of the existing regulatory levels could be identified as the cutoff level, (e.g., the monitoring threshold, see p. 189), or an additional guidance value could be added to 10CRF20. Doses to individuals below the specified cutoff level would be considered of no regulatory concern.

ALARA, "As Low as Reasonably Achievable," Regulation. Current NRC regulations prescribe that in addition to keeping exposures to radiation below the applicable dose limits, licenses must maintain exposures ALARA. There is no lower cutoff dose for ALARA considerations; thus, theoretically, as long as it is *reasonably achievable,* doses should be reduced to miniscule levels. In practice, a cost-effectiveness approach is often taken to decide when to cut off dose-reduction efforts. ICRP encourages this *optimization* technique, and it is used in NRC's Appendix I (see section on p. 183). This approach sets, in effect, a limit on "achievability" defined in economic terms. For example, in Appendix I, dose reduction methods that cost more than $1000 per person-rem saved need not be used.

An alternate approach to truncation of the ALARA process would be, as suggested by Dr. Rossi (see Appendix), establishment of an ALARA regulatory cutoff dose. Doses below that level would be of no regulatory concern, and once that level was reached no further efforts to reduce exposures would be called for. The NRC regulations could set forth ALARA cutoff levels for various activities relating to public and occupational exposures, or could set forth the methodology for determining such levels in specific situations. Since several practices to which ALARA applies might result in small doses to a single individual, the ALARA cutoff level might be set as a fraction of the regulatory cutoff level for individual exposure. The ALARA cutoff could be expressed as a dose or a dose rate depending on the type of practice being considered.

It should be noted that the ALARA principle does not and cannot require the total dose to an individual to fall below the background level. Thus, at some point, the reduced

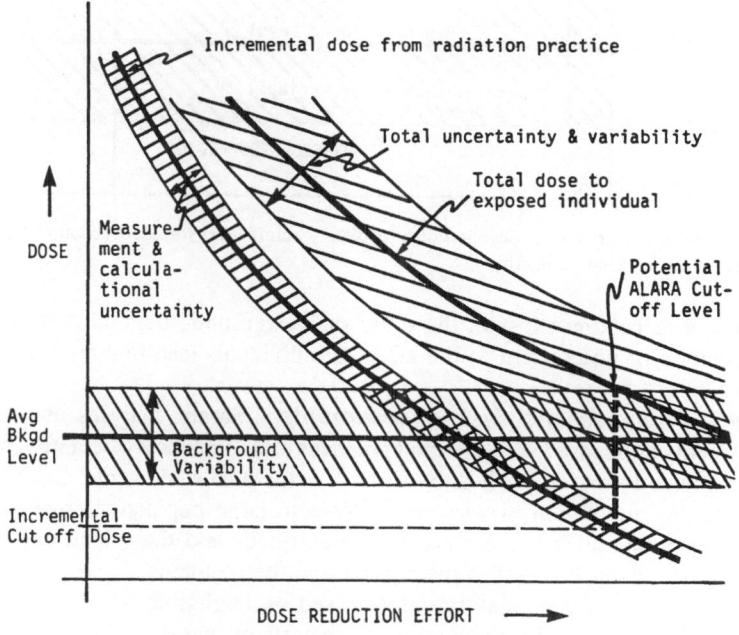

Figure 8. ALARA limited by background.

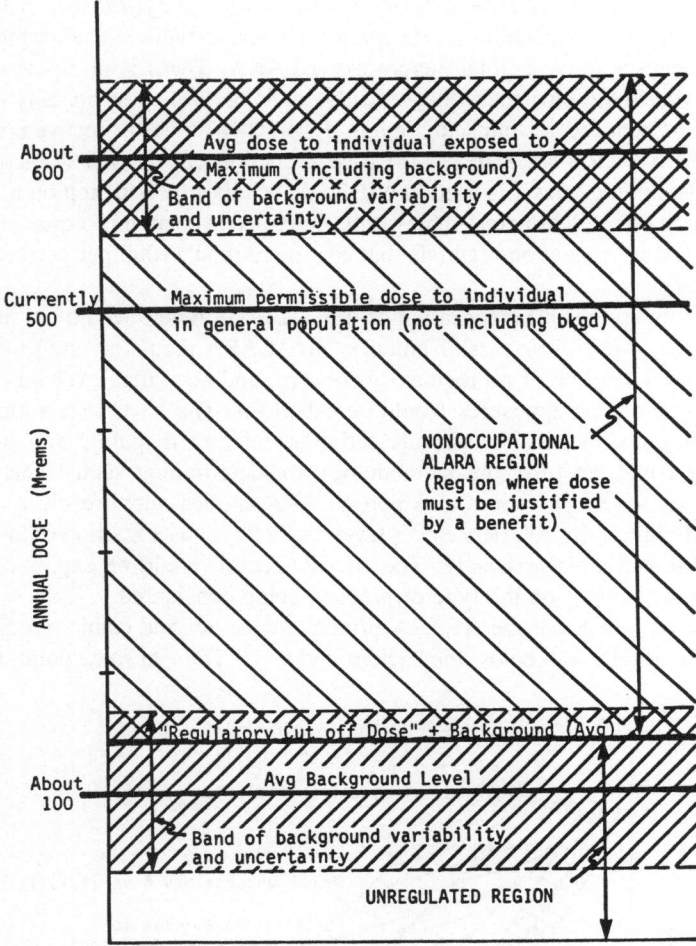

Figure 9. Schematic diagram of application of a regulatory cutoff level for nonoccupationally exposed individuals based on background variability.

incremental dose becomes lost in the *noise* of background. Beyond that point, further reduction can be of no concern to the affected individuals (see Figures 8 and 9).

Safety Analyses and Risk Assessments. Safety analyses and probabilistic risk assessments of licensed activities and practices often include a calculation of estimated impacts, expressed as risks of "health effects," of the analyzed events. Such calculations may involve the summation of very small doses to large populations over long periods of time. In general, no particular *cutoff* is prescribed, and the selection of reasonable limits on time and space is left to the analyst and the regulator.

The usefulness of a regulatory cutoff based on negligible doses in this context is obvious. It would serve to focus attention on the effects that should be of concern, those involving sizable doses to actual individuals.

It should be noted that accident risk assessments generally have two components of health risk; (1) a probability of effects of very low doses, such as those discussed in this report, and (2) a small probability of effects caused by higher doses. The level of risk deemed "negligible" or "acceptable" for these two types of effects may be different. This report addresses only the first type of risk, where the doses of concern are near or below background levels.

The safety analysis cutoff level could be expressed in individual dose or dose rate terms. This individual cutoff dose level could be used to truncate calculations of effects, integration over space and time ceasing when all individual doses have fallen below that level. In probabilistic analyses the risk of deleterious effect would be set equal to zero for any doses below the safety analysis cutoff level.

Again, because of the theoretical possibility of cumulation of doses, the safety analysis cutoff level could be set at a fraction of the individual regulatory cutoff dose.

Environmental Impact Analyses under NEPA. Environmental impact analyses are similar to those described in the previous section, and the applicability of a cutoff level would be similar. In these analyses, however, in addition to, or in lieu of, a cutoff criterion for exposure to an individual, a population (collective) dose cutoff criterion may be appropriate. That is because the environmental impact analyses are performed to evaluate inputs to a cost-benefit balance, although, if impacts are *negligible,* or *remote and speculative,* they need not be considered further in the overall cost-benefit analysis. In some cases it might be deemed necessary to calculate collective effects of doses below the individual regulatory cutoff level to determine the *residual* risk. If the results of such an analysis were a collective dose below a collective cutoff criterion, the effects would be considered negligible and of no further concern to impact assessment. The same effect could be achieved by use of individual dose or dose rate cutoff level that is a fraction of the individual regulatory cutoff dose, as described above.

Waste Disposal Criteria. There is a great deal of activity currently going on in trying to develop limits, often characterized as de minimis, for release of slightly radioactive material to the environment. Often the materials of concern are less radioactive than naturally occurring materials, but, because they arise in a regulated context, they are not easily deregulated. Several areas of current concern are discussed below. Cutoff levels could be used to define the point at which the material becomes of no regulatory concern. This does not mean, however, that in such applications there may not be a higher allowable level set based on cost-benefit or other considerations. The regulatory cutoff levels would be stopping points for the ALARA process as well.

Low-level waste disposal. If a regulatory cutoff policy is adopted, one immediate application would be establishment of levels of slightly radioactive materials that can be discharged to the environment without regulatory concern. This is obviously an area where comparison with naturally occurring radioactive material is appropriate.

One approach to a regulatory cutoff for uncontrolled disposal would be to define material as "radioactive," for regulatory purposes, only if its activity concentration level exceeds that of naturally occurring materials that are deemed to present a negligible risk. The disposal of materials that are not *radioactive* in this sense would not require regulation.

A regulatory cutoff dose level would also have applicability in the design and operation of radioactive waste disposal facilities. If burial sites or other facilities are designed so that there is reasonable assurance that the dose to no off-site individual will exceed the cutoff level, no further dose-reduction efforts need be required, and, other than confirmatory monitoring and record keeping, there would be no regulatory concern.

Long-term aspects of high level and transuranic waste disposal. The evaluation of effects of high level and transuranic waste disposal generally involves consideration of risks over time periods of thousands of years. A regulatory cutoff level would allow truncation of safety and environmental analyses when the radioactive materials concentrations of the waste in question has decayed below a cutoff value—for example, one related to uranium ore radioactivity. The negligible risk posed by the material thereafter would be of no regulatory concern.

Similarly, in the design and operation of waste storage facilities, the cutoff dose level would provide a stopping point for ALARA considerations.

Uranium mill tailings control. The tailings materials from the mining and milling of uranium naturally emit radon gas. When these materials are stored on the ground surface, or used for construction or in other applications, radioactive gas that would otherwise have remained within the earth can enter the atmosphere and increase the dose from radon gas and its radioactive daughter products to human populations.

A regulatory cutoff policy would help define the conditions under which tailings materials are of no regulatory concern. This is another area where comparison with naturally occurring sources would be appropriate. Regulatory cutoff levels for radon emanation rate from materials, activity concentration within materials, and dose rate to individuals, could be established, on the basis of comparison with appropriate background levels. Materials that satisfied all cutoff criteria would be considered nonhazardous, and no further efforts would have to be devoted to their control.

Nuclear facilities decommissioning. Like waste disposal, decommissioning requires definition of when a *radioactive* component or material can be released for disposal or reuse. A regulatory cutoff level, in this context, could be used to define the level at which a scrap material ceases to be considered *radioactive* and becomes just another piece of steel, copper, etc. Material that presents a negligible risk to the public could then be treated the same as any other scrap from a dismantlement project.

Another application would be decommissioned sites. If the doses to the public from radioactive materials that remain at a decommissioned site are kept to level below the regulatory cutoff, there would be no further regulatory concern, and no efforts would have to be made to reduce them further pursuant to ALARA.

Decontamination after accidents. After a contamination accident, decontamination may be necessary. There is always a question of the level to which decontamination must proceed. If a regulatory cutoff policy is adopted, the contaminated area could certainly be deemed *clean* if its activity were less than the cutoff level. However, on a cost-benefit or other basis, decontamination only to a certain higher radiation level ("clean enough") might be deemed appropriate in many cases. In those cases the ALARA principle would apply, but only down to the cutoff level.

Appendix I to 10CFR Part 50 and Environmental Technical Specifications.
Appendix I to NRC's regulations, 10CFR Part 50, provides numerical guides for reactor effluents. The current guide were set on the basis of the industry's ability to reduce calculated off-site dose rates by rigorous design of radioactive waste control systems. It was deemed "reasonably achievable" for incremental whole body doses to the maximally exposed individual, calculated using conservative models, to be kept below 3 mrems per year from liquid effluents and 5 mrems per year from gaseous effluents. In addition, individual facilities are required to lower these doses further, if such reduction can be achieved at a cost of less than $1000 per person-rem saved.

If the NRC adopts a regulatory cutoff policy, the Appendix I levels might have to be changed to be consistent with that policy. The doses set forth as guidance in the Appendix are at or below the levels that many experts would characterize as de minimis, yet they were not set as limits below which there is no regulatory concern, since in some instances lower levels may be required.

As observers have noted,[54] the guidance levels of Appendix I are set too low to be generally detectable above background. Compliance is achieved, at both the design and operational stages, through calculations of dose based on mathematical models. Thus, if regulatory cutoff levels are established on the basis of detectability of radiation levels in background fields or comparison to variations in background, the dose levels of Appendix I would probably have to be increased, and the requirements to go to even lower doses in some cases could be eliminated. The levels of a revised Appendix I could be set at the individual regulatory cutoff level or somewhat higher, reflecting that a risk somewhat above a negligible level might be acceptable because of the offsetting benefit of the use of nuclear generated electric power.

To ensure compliance with Appendix I, the NRC is promulgating environmental technical specifications (ETS), limiting emissions, and requiring environmental monitoring programs of licenses.[56] If comparison with the regulatory cutoff limits indicates that the risks associated with Appendix I guidance dose levels are negligible, the need for extensive environmental technical specifications may be eliminated (see p. 184).

Safety Goals and Siting Policy. The NRC's work on safety goals and siting policy currently involves analyses of the impacts on surrounding populations caused by routine releases and accidents at licensed facilities. Insofar as they consider effects of low level radiation, such analyses could be affected by a regulatory cutoff policy, as discussed previously.

When the NRC sets a *safety goal,* of whatever form, it should take into account the fact that the risk of detrimental effects from radiation at the previously set individual regulatory cutoff level has been determined to be negligible. An allowable risk level or safety goal should logically be set at a value above the cutoff risk level for low doses of radiation, since the allowable risk includes accidental risks and involves consideration of offsetting benefits (see Figure 10).

Other Regulations. NRC regulations in other areas (e.g., transportation packaging) may be affected by a regulatory cutoff policy. The ALARA aspects of the policy will apply to all areas of radiation exposure regulation.

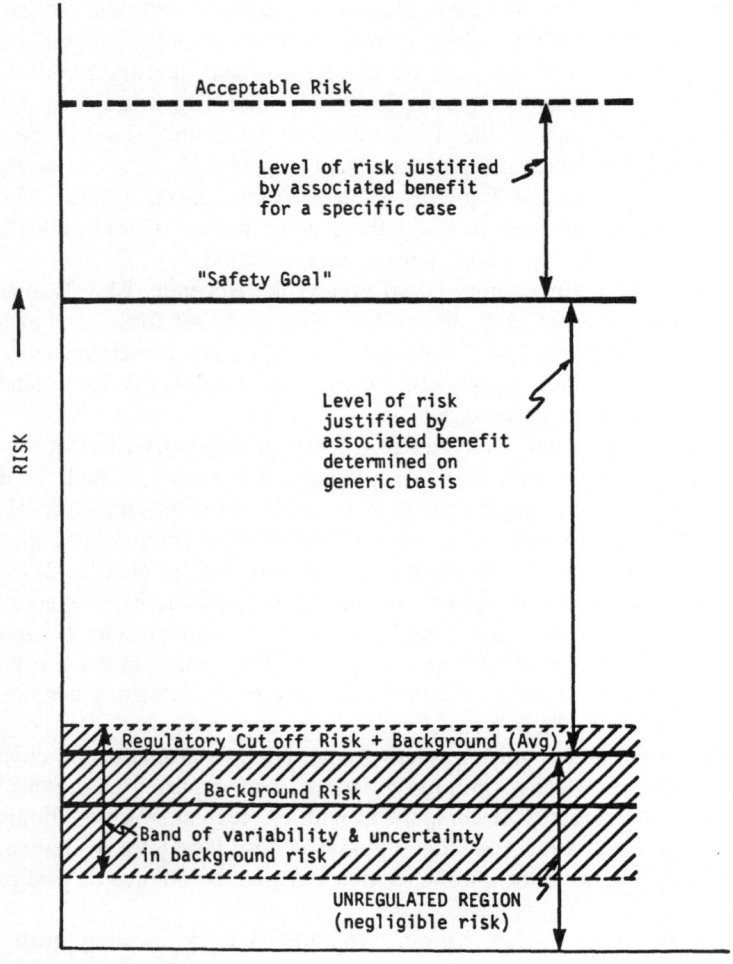

Figure 10. Schematic diagram of relationship of safety goal to regulatory cutoff risk.

Regulatory Policy. An area in which a regulatory cutoff policy would be extremely useful is in setting priorities for standards development, promulgation of regulations, and enforcement actions. Agency actions that would reduce radiation exposures that are already at or around the regulatory cutoff or even lower levels would be seen to be essentially meaningless from a health protection standpoint. Similarly, actions that would change exposures by an amount on the order of the cutoff levels would be difficult to justify. As a result, regulatory concern and resources could be focused on those areas where significant health and safety problems exist. A population (person-rem) dose cutoff criterion might prove useful in this context as well.

For example, the NRC is now attempting to implement radiological environmental technical specifications requirements for generating plants. These are concerned mostly with Appendix I limits. Even without these specifications, the NRC staff admits, "the general objective of the effluent control program evidently is being met."[56] The annual collective dose to the U.S. population from all nuclear power plants combined is now about 150 person-rems. The average dose to an individual in the exposed population is calculated to be less than 0.008% of the natural background radiation level. Of course, this incremental dose cannot be detected in the variability of background. If there were a regulatory cutoff policy for individual dose or population dose, the levels cited above would be below the cutoff level set on any reasonable basis. The implication would be that these effluent doses are of no regulatory concern and the agency should direct the effort being spent on radiological environmental technical specification development to work that is more productive of health and safety benefits.

Applicability of Cutoff Levels to Other Agencies

Adoption of a regulatory cutoff policy for radiation exposure by the Environmental Protection Agency, acting in its role as developer of radiation protection guidance for all federal agencies, would affect not only those federal agencies directly but the state and local regulatory agencies that model their regulations on the federal ones, as well.

Absent such a general governmental policy, individual agencies might establish such a policy on their own. In addition to NRC, discussed above, these included the Department of Energy (regulation pursuant to UMTRCA, regulation or research activities in National Laboratories), the Environmental Protection Agency (regulation pursuant to Atomic Energy Act, Clean Air Act, UMTRCA, etc.), and the Food and Drug Administration (control of consumer products).

State agencies and institutional entities such as the Institute of Nuclear Power Operations (INPO) might also establish low-dose cutoff policies.

Applicability of Radiation Dose Cutoff Levels in Other Areas of Public Concern

The adoption of a regulatory cutoff policy for radiation exposure by one or more federal administrative agencies would serve as guidance and precedent for the Congress, the states, the private sector, and the courts, and as information for the communications media and the public.

The Congress, for example, is currently considering legislation to compensate persons who were exposed to radiation from weapons tests and may have been injured as a result. A de minimis approach might be used to determine areas where the risks of injury were negligible and therefore compensation is unjustified.

The courts have, in some radiation cases, adopted, in effect, a de minimis approach. For example, in *Peshlakai* v. *Duncan* (an NEPA case relating to a proposed *in situ* uranium leaching project), the court compared the radon release levels to background levels and found the radiation threat "realistically nonexistent."[57]

On the other hand, in the Silkwood case[58] (a tort case alleging injury from plutonium contamination), the district court apparently endorsed the proposition that exposure to any amount of radiation could cause compensable injury, because absorption of any radiation causes some disruption to components of body cells. This rationale is certainly incompatible with a de minimis approach. It not only says that any incremental dose of radiation can cause later harm, it says, in effect, that, as matter of law, any incremental dose of radiation does cause immediately compensable injury. The existence of a well-articulated regulatory cutoff policy and its presentation to the court might result in a different result in a similar case in the future.

Bases and Methodologies for Setting "Regulatory Cutoff"

This section presumes that an organization has decided to adopt a *regulatory cutoff* policy. It describes some of the ways in which the regulatory cutoff level(s) could be derived. For the purpose of discussion in this section, we shall concentrate on whole-body external doses to individuals in the public. The concepts involved can readily be extended to other types of radiation exposure (e.g., dose from internal emitters). Also, for discussion purposes, the NRC is generally taken as the regulatory agency setting the levels. Numerical values based on current and suggested levels are summarized in Table 7.

Derivation from Current NRC Standards. One way in which *regulatory cutoff* levels might be set is by derivation from existing regulatory levels of individual dose or from the individual risks such dose levels are estimated to represent. To do so, one must examine the expressed or implied bases of such levels, to determine if they are consistent with the rationale of a *regulatory cutoff*, that is, a level of radiation that is of no regulatory concern.

Current NRC regulations, 10CFR Part 20, include permissible doses, levels, and concentrations of radiation and radioactive materials. For example, in 10CFR20.105, licensees are permitted to expose members of the public in *unrestricted areas* to radiation from licensed sources if it is "not likely to cause any individual to receive a dose to the whole body in any period of one calendar year in excess of 0.5 rem" (500 mrems).

The promulgation of such a permissible level by a regulatory agency might be considered to imply that exposure to radiation at this level by the agency represents *no undue risk* to the public health or welfare. It should be noted that the voluntary standard-setting bodies, NCRP and ICRP, have also recommended 500 mrems per year as the maximum permissible dose (mpd) limit for the public.

Maximum permissible doses. A simple proposal for a regulatory cutoff level for public exposure would be to set the cutoff level at the current mpd level. This would be close to *deregulation*. It would leave to the licensees the entire task of determining how to control radiation sources and controlling them so as to prevent exceeding the limit. At least one health physics expert has proposed a possible de minimis radiation level of about 500 mrems.[2]

There are some conceptual objections to declaring doses of the order of the current mpd limit to be *of no regulatory concern.* First of all, the current regulations include [Section 20.1(b)] in addition to the dose limits, the statement that licensees should "make every reasonable effort to maintain radiation exposures . . . as low as is reasonably achievable." Use of this so-called ALARA policy, which has not been defined with any explicit lower limit, has also been recommended by the standard-setting bodies. Pro-

Table 7. Numerical Dose Values—Possible Bases for Regulatory Cutoff Levels

Application	Cutoff level	Notes	Reference
Individual dose limit (general population)	500 mrems/yr		NRC 10CFR20
Grand Junction, Co.; cutoff for remedial action	440 mrems/yr	Based on 0.05 mrems/hr external radiation	DOE 10CFR712
Monitoring threshold for radiation workers	1200 mrems/yr	Value applicable to adult worker	NRC 10CFR20
	250 mrems/yr	Value applicable to minor worker (member of the public)	
Indoor contamination cutoff for remedial action cleanup of inactive uranium mill sites	180 mrems/yr (including background)	0.02 mrems/hr (including background indoor gamma radiation)	EPA 40CFR192
Regulatory cutoff level—negligible dose	100 mrems/yr	Risk of 10^{-5} correlatd with dose based on conservative estimate of 10^{-4} fatal cancers per rem	C. Comar, *Science* 203, p. 4378
Population exposure ALARA guidance for NRC licensees	100 mrems/yr (proposed)	Between this level and the 500 mrems/yr limit, ALARA program would be required	NRC staff draft revision 10CFR20 (12/81)
Guidance for disposal or storage of uranium and thorium from past operations	90 mrems/yr	0.01 mrems/hr (over background) proposed for inactive uranium processing plants	NRC branch technical position
Dose limits (general population)	25 mrems/yr	Applies to public exposures from uranium fuel cycle activities	EPA regulations 40CFR190
Contamination level limit, decommissioning	25 mrems/yr (suggested)	One-fourth of average background; based on detectability in varying background	ORNL report by Holoway et al.— NUREG CR-2082
Population exposure cutoff for NRC licensees	25 mrems/yr (proposed)		NRC staff draft revisions to 10CFR20 (12/81)

(continued)

Table 7. (Continued)

Application	Cutoff level	Notes	Reference
Dose from mill tailings disposal	14 mrems/yr	Dose at 100 m from tailings pile; based on 2 pCi/m^{2}sec emanation rate of radon gas	Final generic EIS on uranium milling, NUREG-0706
Limits on exposure from reactor effluents (general population)	8 mrems/yr	Total dose from air and water effluents in the environment	NRC regulations 10CFR50, Appendix I
Nuclear facilities decommissioning residual radioactivity	5 mrems/yr (proposed)	Staff-based value on consistency with existing guidance	NRC staff report NUREG-0613
Nuclear facilities decommissioning residual radioactivity	1 mrem/yr (proposed)	Value raised for comment	NRC staff report NUREG-0436 Rev. 1
Allowable dose from solid waste requiring no regulatory control or surveillance	1 mrem/yr (suggested)	Based on comparison with NRC 10CFR50, App. I and 10CFR20	AIF study (1978)
Dose with no significant impact	0.2 mrems/yr	17 person-rems/yr from plant compared to 73,000 from natural background	NRC Appeal Board Maine Yankee, ALAB-161 ALAB-175
Cutoff dose for environmental analyses	0.1 mrems/yr (proposed)	Based on conservative dose estimate for 10^{-8}/yr	NRC staff draft revisions to 10CFR20 (12/81)
De minimis level for NEPA cost/benefit balancing	0.02 mrems/yr (average over population)	Based on 0.5 death per year in population of 300 million as a "minimal impact"	NRC Licensing Board (Perkins Plant) LBP-78-25
Cutoff level for regulatory concern	Somewhere between 0.01 mrems and 10 rems	Based on a risk of 10^{-9} being obviously negligible, and 10^{-3} being of possible concern	U.S. Supreme Court (Benzene case)

mulgation of the ALARA policy seems to imply that there is some risk associated with doses at the regulatory limits and the risk should be reduced further where possible. This is consistent with the rationale of a standard set on the basis of balancing risks and benefits. Thus, the mpd level might be viewed as the level at which the societal risks of the radiation exposure are balanced by the benefits of the use of the licensed sources. On that basis the mpd level would not necessarily be a level that is de minimis in the technical sense or "of no concern" in the regulatory sense (see Figure 10).

Another objection to the use of the mpd as the regulatory cutoff level is the possibility that individuals could receive mpd doses of radiation from several sources, cumulating to levels that are of regulatory concern. This problem, which is often a logical rather than a practical one, exists with the current regulations as well. It is discussed further in the section on development of a regulatory cutoff policy.

Unmonitored exposures.   NRC regulations (Section 20.202) relating to personnel monitoring provide another basis for a regulatory cutoff. Under that section, the dose to people in restricted areas need not be monitored unless it is likely to exceed about 300 mrems per calendar quarter for an adult, about 60 mrems per quarter for a minor. It appears that the NRC considers occupational doses below such levels to be of little or no regulatory concern since no records of such doses are required to be made or kept. On this basis, these levels might provide appropriate *regulatory cutoff* levels, the one for adults covering adult radiation workers, the one for minors covering minor workers, and the one for the general public which contains individuals of all ages. If the current NRC values were adopted, that would mean a regulatory cutoff level for an adult of about 1200 mrems per year and a cutoff level for an individual in the general public of about 250 mrems per year.

ALARA levels for reactor effluents.   Appendix I to NRC regulations, 10CFR50, provides numerical guides for design objectives and limiting conditions for operation to meet the ALARA criterion for radioactive effluents from power reactors. Regulatory cutoff levels could certainly be derived from the Appendix I levels. Indeed, as described in the section on current proposals, a contractor for the Atomic Industrial Forum (AIF) reviewed current radiation regulations and determined that to be compatible with current governmental standards and regulations, a "de minimus" (*sic*) dose rate, applicable to uncontrolled disposal of soil radwaste, must be less than the most restrictive levels, the *guidance* levels of Appendix I. The report's authors further reasoned that since the Appendix I levels are required to be monitored, they cannot be considered totally negligible; an additional *factor of disregard* was therefore applied to them to determine a regulatory cutoff level. This factor of disregard was deduced to be 25%. Applying this 25% to the 3 to 5 mrems per year of Appendix I, the AIF report authors proposed 1 mrem/year as their "de minimus" level of dose.

Comparison to other regulated levels.   A *relative* regulatory cutoff level could be derived by defining as *negligible* levels that are small compared to regulated levels. This is analogous to the relative de minimis approach described earlier. Such an approach may be useful at least in cutting off the very low dose *tail* in collective dose evaluation over space and time. The determination of what is relatively *small* could be based on little more than the ordinary meaning of the word as interpreted by the agency—i.e., less than half, a few percent, less than 1%, etc.

Other regulatory limits or guidance values. As discussed on p. 169, there are other NRC regulatory limits, exempt and unimportant quantities, and guidance levels set by other agencies, that might be used singly or in combination as a basis for establishing a regulatory cutoff dose. It is incumbent upon the person or agency doing so to specifically set out the rationale for the selection made.

Derivation from a "Safety Goal." NRC has announced its intention to consider the feasibility of establishing a *safety goal* for regulation.[59] For example, this goal might be expressed quantitatively as an annual individual risk value or as a measure of societal risk. A variety of bases for setting such values are being considered, and many of the bases for such values are related to those that are discussed in this report as possible bases for de minimis or regulatory cutoff levels. Ideally, the rationale used to determine the risk level should be compatible with that used to determine the cutoff dose.

If a safety goal were determined before a cutoff dose level, it should then be possible to derive a cutoff level for NRC regulatory applications from that safety goal. For example, if the safety goal were expressed as the individual risk that is deemed *acceptable* in light of the benefits of the use of radiation sources, the negligible risk level for radiation might be established as a small fraction of that risk (see p. 191).

It appears that the establishment of a safety goal will take some time. The subject is more complex than that of low-level radiation risk alone, since it includes accident risks and such considerations as societal risk-aversion and equity, among others. A regulatory cutoff level for radiation could be more easily and quickly determined, and while a safety goal cannot be deduced necessarily from a cutoff dose or risk level, the considerations that go into setting such a level should illuminate many of the factors that must be taken into account in setting a safety goal. Eventually, after a safety goal is established, a previously established individual dose regulatory cutoff level could be modified, if necessary, to be consistent with the newly expressed safety goal (see Figure 10).

Derivation from Expert Opinion. If it is desired to base the determination of regulatory cutoff level for dose, risk, or health effect as much as possible on objective scientific information, the help of experts in the radiological sciences and engineering must be sought. Incorporating expert opinion into the regulatory rule-making process might be done in various ways. However the final decision on the policy and its implementation is the agency's.

Adoption of recommendations of expert bodies. The regulatory agency can defer to the expertise of such bodies as NCRP and adopt the recommendations made by that body as agency policy. Probably additional work would have to done by experts on the agency staff to apply the guidance to specific regulatory situations and develop regulations. At the present time no expert body has made specific recommendations concerning de minimis levels, although NCRP Committee 1 is considering the matter. Other bodies that develop consensus standards might also offer input to a rulemaking on adoption of a de minimis policy and its regulatory applications. They include the American

Nuclear Society (ANS), the Institute of Nuclear Power Operations (INPO), and the Health Physics Society (HPS), among others.

Utilization of agency staff expertise. The experts on the agency staff may themselves develop guidance for a regulatory cutoff policy based on de minimis doses determined by one or more of the methodologies discussed under The Technical Context. Their proposals would then be the basis for proposed regulations upon which other experts, as well as the general public, might comment in a rule-making proceeding.

Determination of collective expert opinion. Often experts differ on the best methodology to use to achieve a derived result and the best information to use as input to that method. In determining de minimis levels, it is quite likely that a divergence of expert views will be seen. Semiquantitative methods exist for drawing reasonable conclusions from varying expert opinions. These range from use of the *most conservative* value recommended by any expert, through simple averaging of recommended values, to such sophisticated methods as the *delphi* approach, to try to get a probability distribution over possible values. The results of this process could then form the technical basis for a rule-making proceeding.

Probabilistic decision analysis. Decision analysis uses a graphic expression of the elements of a decision (decision tree) and estimates of the probability of various outcomes to determine the optimum action. This method might be used, for example, in quantifying the opinions of experts and in combining the opinions of several experts. Further discussion of this technique is beyond the scope of this chapter.

Derivation from Dose, Effect, or Risk Levels That Are Deemed "Acceptable" to the Public. Certainly a radiation level that is of no regulatory concern (i.e., the regulatory cutoff level) must be, at the very least, a level that would be acceptable to a well-informed, reasonable public.

One way to determine what levels of risk (the concept is generally expressed in terms of risk that must be correlated with dose in this case) society finds acceptable is to determine what risks society currently accepts. It is presumed that if people or their governments do not take action to reduce risks, then those risks are *accepted,* and other risks that are qualitatively and quantitatively similar would be *acceptable*. At the present time there is a great deal of interest in this area, especially related to such applications as determination of a safety goal and probabilistic risk assessments for nuclear power plants.

While an existing known risk that is of no concern must be *acceptable,* the converse is not necessarily true. That a risk is *acceptable* does not necessarily imply it is *negligible*. For example, a substantial risk may be deemed *acceptable* if it is outweighted by a greater benefit. The risk level for regulatory cutoff therefore, should probably be defined as "negligible" compared to *acceptable* risks.[26] Variations or uncertainties in acceptable risks might form a basis for evaluating negligibility, or an arbitrarily determined *small fraction* could be used. Sociological studies might provide information on what levels and types of risks are treated as negligible by our society.

As noted in the Appendix, Drs. Webb and McLean used 10^{-6} as a *negligible* annual level of risk. Dr. Comar (p. 177) used 10^{-5} as his estimate of negligible risk, but only if there is some benefit associated with it and it is not easily avoided.

There is some controversy today over whether the qualitative aspects of risk should be taken into account. Two risks of equal numerical value may evoke different public responses. One of them may be subjectively perceived to be less *acceptable* because of its dread nature, its unfamiliarity, or other factors. This concept of *perceived risk* may become especially important in the regulatory field. There may be so much public and congressional concern that the agency must respond by regulating a risk that might objectively be considered *negligible*.

A final problem is that of who decides what is *acceptable*. Is it a subject for expert testimony or a public policy decision to be made by the agency?

Limitation to Less Than One Projected Health Effect. Setting, as the regulatory cutoff level, the radiation dose that causes not even one health effect in a population might be an attractive option for regulators. There are several problems in establishing such a limit. First of all, an appropriate dose–effect relationship must be used. The regulator must decide how *conservative* a relationship to assume. Second, the level of dose determined depends on the size of the population affected. Thus, for each situation evaluated, the individual cutoff dose might be different. For example, if it is assumed that 10,000 person-rems produce 1 health effect (individual risk of 10^{-4} per rem), then considering doses to a population of 1 million, the regulatory cutoff dose would have to be less than 10,000/1,000,000 or 10 millirems. If it were desired to establish a cutoff dose level for general applicability to radiation uses that might affect the entire U.S. population (about 250,000,000 people), the de minimis dose, using the same dose–effect assumption, would be about 0.04 mrems.

A similar analysis can be used to establish a regulatory cutoff collective dose (collective dose can be considered a measure of societal rather than individual risk). For example, given the conservative dose–effect assumption cited above, a collective dose cutoff might be set at 10,000 person-rems. Any practice with a lower collective dose could be considered to be of negligible effect, subject, of course, to the limitation that within the population in question no exposed individuals are exposed to doses in excess of the individual limit.

This methodology could be generalized to be based on the calculation of an estimated number of health effects deemed *negligible* by society. This number might be greater, less, or much less, than one.

Determination by Risk-Cost-Benefit Balancing and Trade-Offs. Risk-cost-benefit balancing and trade-offs are rationales for selection of regulatory cutoff levels that may also involve others of the methodologies discussed in this section. They are characterized by the attempt to compare and balance things expressed in disparate units (dollars, lives, mrems) and to bring them to a common basis for decision making. Several conceptual approaches are discussed below.

Negligible benefit of reduction of residual risk to zero. The cutoff level can be considered to be one where there would be a *negligible* benefit from further reduction of exposures all the way to zero. The benefit, reduction in health effects, could be evaluated on the basis of a dose–effect model of the type described earlier in this report. If the linear, nonthreshold, dose-rate-independent extrapolation were used, the result would be a *conservative* overestimate of the benefit. The *negligibility* of the dose, risk, or effect would be determined on the basis of that due to background, that due to other accepted risks, or the number of health effects deemed acceptable (e.g., less than one).

Comparison with competing methods for achieving a benefit equivalent to that of a reduction of residual risk to zero. The regulatory cutoff level can also be considered to be one where the expected benefits (expressed in terms of health effects) for reduction of exposures all the way to zero could be more readily, efficiently, or cheaply achieved for society in some other way, and society has chosen not to do so. The inference is that society has decided that such a level of effects is of no concern.

Cost-benefit balance expressed in monetary terms. Another approach is to express the value of the lives that are estimated to be saved by reduction of the cutoff dose to zero in terms of dollars. This is difficult to do particularly in a regulatory rule-making format. While governmental decisions must often be based on implied values for a human life, and studies have made these values explicit, public acceptance of a method that is based on a stated evaluation that, say, one life is worth $100,000 would be hard to achieve. In any event, once a dollar value is determined, it must be compared to some other economic parameters—for example, the cost of achieving the postulated reduction. The level at which the calculated cost exceeds the calculated benefit, assuming it is below the individual dose limit, could be taken as the cutoff level. Alternatively, a level that is a fraction of the level at which cost and benefit balance could be established as the regulatory cutoff. It should be noted that when radiation effects are to be balanced or compared to economic or other parameters, a realistic estimate of the dose–effect relationship, rather than a conservative estimate, should be used.

Cost-effectiveness. A *cost-effectiveness* criterion can also be used to determine a regulatory cutoff level. In this methodology, the cost for each increment in dose, effect, or risk reduction below the limit is determined. As the level is reduced below the mpd, the cost per unit level of reduction increases. Often an inflection point is observed in the curve of cost versus effect (see Figure 11) beyond which very small relative reductions in effect require large increases in cost. The level of the inflection point, or some fraction of it, can be taken as the regulatory cutoff.

Guidance for Regulatory Decision Making

Legislation. A principal source of guidance for a regulatory agency comes from Congress in the form of legislation and annual appropriations. The current statutes under which ionizing radiation is regulated are discussed on p. 169.

Congress could adopt legislation explicitly establishing a policy of regulatory disregard for negligible levels of radiation exposure. Such legislation could range from a simple declaration of policy, with details to be worked out by the agency, to a detailed specification of the various applications of regulatory cutoff policy, the value of the de minimis dose or risk, etc., with little left to agency discretion. The former would probably be the most reasonable approach.

At the present time, no legislation on a regulatory cutoff policy applicable to ionizing radiation has been proposed in Congress.

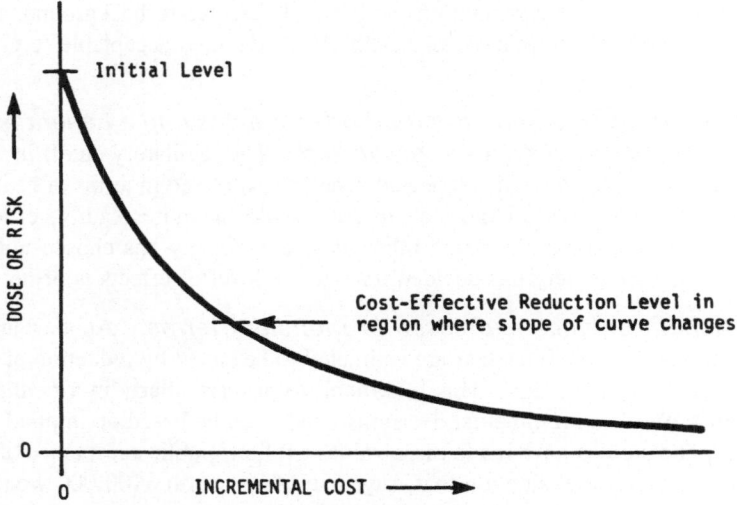

Figure 11. Cost-effectiveness in dose or risk reduction.

Administrative Guidance. The establishment of an administrative policy for a regulatory cutoff for negligible radiation doses could be issued by the office of the president. Guidance might also emanate from EPA, which retains the authority, formerly held by the Federal Radiation Council, to establish guidance for federal agencies in the area of radiological health and safety. A policy might also emanate from the surgeon general (Department of HHS), who is charged with protection of the nation's health, or the Council on Environmental Quality (CEQ), which provides guidance on environmental matters for federal agencies.

Guidance by the Courts. The courts of the United States and of the states decide controversies between parties and, in the process, interpret statutes and regulations. Those decisions establish precedents and influence subsequent agency actions. While court decisions cannot actually establish policy, they can interpret legislation or administrative policies under a variety of circumstances and profoundly effect an agency's interpretation of its mandate.

It is beyond the scope of this report to review or discuss the many court precedents that might be cited in support or against the reasonableness of a de minimis dose or *regulatory cutoff* policy under current laws (see footnote on p. 146). It should be noted, however, that the U.S. Supreme Court, in the so-called benzene case,[60] recognized the existence of *insignificant* risks. In so doing, it offered quantitative bounds as follows: "If, for example, the odds are one in a billion that a person will die from cancer by taking a drink of chlorinated water, the risk clearly could not be considered significant. On the other hand, if the odds are one in a thousand that regular inhalation of gasoline vapors which are two percent benzene would be fatal, a reasonable person might well consider the risk significant and take appropriate steps to decrease or eliminate it." In his concurring opinion, Chief Justice Burger stated: "Inherent in this statutory scheme is authority to refrain from regulation of insignificant or de minimis risks." See [*Alabama Power Com-*

pany v. *EPA* (D.C. Cir., Dec. 14, 1979).] When the administrative record reveals only scant or minimal risk of material health impairment, responsible administration calls for avoidance of extravagant, comprehensive regulation. Perfect safety is a chimera; regulation must not strangle human activity in the search for the impossible.

Guidance by Expert Bodies.   As discussed elsewhere in this report, the expert bodies, such as NCRP and ICRP, that are active in the field of radiation protection, are available to provide guidance for regulatory agencies. Other such bodies include the National Academy of Sciences (including its BEIR committee), the National Academy of Engineering, the International Atomic Energy Agency, the United National Scientific Committee on the Effects of Atomic Radiation (UNSCEAR), the American Nuclear Society, the Health Physics Society, and the American Physics Society. As noted in this report, some of these bodies have been actively considering de minimis approaches.

Guidance by the Public.   The regulatory agencies are ultimately responsible to the public, primarily through its elected representatives in Congress and the executive department. The public, or, at least, organized components of it, also participates in agency rulemaking and makes its influence felt through the media of public communication.

Among the groups of the public that may influence regulatory agency actions are (a) industry and regulatees, (b) special-interest groups, such as environmental, pro- and antinuclear, and consumer organizations, and (c) members of the public or groups who have particular or local concerns (e.g., those who live near a proposed licensed facility).

The agency rule—making procedures generally provide a forum for these groups to present their positions. If anyone is dissatisfied with an agency decision, recourse to the courts is possible.

Such groups can also propose agency action. Thus, for example, a group of its licensees could petition NRC for a rule making on a proposed application of de minimis levels in a regulatory cutoff policy.

DEVELOPMENT OF A REGULATORY CUTOFF POLICY

Policy Statements

Federal Policy.   The most reasonable approach to a regulatory cutoff policy for radiation exposure would be the development of a policy to cover all federal government activities. Such a policy could be promulgated by the president, with guidance from the EPA and perhaps the CEQ. Individual federal agencies would then apply the policy to their regulatory activities and develop specific regulations.

NRC Policy.   The Nuclear Regulatory Commission need not wait for a general governmental policy on regulatory cutoff levels to be developed. The agency has authority under the Atomic Energy Act to regulate the use of source, by-product, and special nuclear materials to protect the public health and safety. The adoption of a cutoff policy for radiation exposure would be consistent with the agency's mandate.

Presentation of Proposals for a Regulatory Cutoff Policy

Presentation to the Executive Branch. Proposal for a general federal regulatory cutoff policy for radiation exposure could be made directly to the office of the president through the president's science advisor or of the Council on Environmental Quality. The Environmental Protection Agency, as the organization responsible for promulgation of federal guidance, could also be approached, for example through a petition to the administrator.

A proposal for a regulatory cutoff policy for the Nuclear Regulatory Commission could be made by special petition to the commission, requesting development of a policy statement; by a formal petition for rule-making, asking for revision of all applicable NRC regulations to reflect a regulatory cutoff approach; or by formal petitions for rule-making in specific areas where such a cutoff level would be of immediate and significant applicability, such as Appendix I to 10CFR Part 50. An opportunity for adoption of a cutoff policy will be provided by the planned NRC rule making on revision of the commission's radiation exposure regulations, 10CFR Part 20. That proceeding is likely to provide a forum for presentation of a proposal for such a policy and development of related regulations.

Presentation to the Legislative Branch. The Congress could legislate a regulatory cutoff policy for radiation regulation applicable to all federal activities or to specific agencies. The preparation of draft legislation on this matter for presentation to the staffs of the appropriate congressional committees would be a major first step toward the legislation of such a policy.

Presentation to the Public. A major benefit of the adoption of a regulatory cutoff policy for radiation exposure is expected to be an increased public confidence in the regulatory program. For this to occur, however, the public must be informed about the policy, its basis, and its effect. Therefore, a program of information and education, for the public and for general information media representatives, must be an essential part of the development of this policy. While some of this education can be done by the federal agencies, there is a large role for NRC licensees, other industry groups, and the scientific community.

Another audience that must be considered is state officials with regulatory and legislative roles. Much regulation of radiation exposure is carried out on the state level, and it is essential that the policies of the state and federal entities involved be consistent. This concept could be presented as representing levels that are SETE, "safe enough to exempt," from regulatory control.[3]

PROBLEMS AND BENEFITS OF A REGULATORY CUTOFF POLICY

Potential Problems Involved in Adoption of a Regulatory Cutoff Level for Radiation Exposure

This section discusses the potential problems that might arise in the development, adoption, and application of a regulatory cutoff policy. The benefits associated with the implementation of such a policy are addressed on p. 198.

Opposition.   The first problem with the proposal for a cutoff policy arises from the probability of strong and vocal opposition to such a policy from those who are opposed to the use of nuclear energy or have strong fears of radiation. Both the proponents and the opponents of the policy will have to expend time, resources, and expertise in supporting their position in agency and court proceedings. Governmental agencies will also have to devote resources to the development of the policy and the holding of public proceedings.

The proceedings are likely to attract media attention to areas of conflict. As a result, the public may become even more confused about the risks of radiation exposure, and further fears may be aroused.

To minimize this problem, the proponents of a cutoff policy must have well-reasoned and scientifically valid positions and must articulate them in a way that is understandable to members of the public, their governmental representatives, and the media.

Regulatory "Ratcheting."   Another problem inherent in adoption of any numerical guidance for regulatory cutoff is the tendency for the levels set to become upper limits instead of guidelines for negligibility. Regulatory agencies may find it easier to presume that the cutoff levels are *reasonably achievable* and require everyone to conform to these levels in place of the existing higher maximum permissible levels.

The best way to minimize this problem is probably to state the policy in specific terms that make clear that the de minimis levels are not maximum permissible levels, and strongly challenge any agency interpretations to the contrary.

Accumulation to Significant Levels.   A major conceptual problem with the use of a regulatory cutoff level is the possibility that a particular individual could receive doses at or near the cutoff level from several sources. If enough *negligible* doses are added together, the result would eventually be significant. For example, if the de minimis level for the maximally exposed person were set at 100 mrems per year, a person receiving that dose from 50 different sources in a year would accumulate an annual dose of 5,000 mrems (5 rems), the average occupational dose limit. In practice, of course, the chance of such an occurrence is very small. For example, few, if any, members of the general public could be the maximally exposed to radiation from more than one major NRC-licensed facility for geographical reasons.

The solution to this perceived problem is twofold. First, the agency should adopt a tiered system of regulatory cutoff—that is, the de minimis dose rate should be the first guide for the agency. The agency should use that level as a basis for setting regulatory cutoff levels for individual licensees. A licensee, in turn, could set cutoff limits for individual practices that would make it unlikely that any person would be maximally exposed to incremental radiation from so many practices that his annual dose would substantially exceed the licensee's cutoff level. So, for example, a nuclear power plant facility that conducts ALARA reviews for many practices might set an ALARA cutoff level at 1% of the facility individual dose regulatory cutoff level.

Second, the agency should monitor environmental radiation levels to identify long-term trends. The de minimis doses and regulatory cutoff levels should be reviewed on a periodic basis, say every 10 to 15 years, to check their validity against then current radiobiological information, and to adjust the cutoff levels, if necessary, to ensure that significant long-term increases in background levels due to the cutoff policy do not occur.

Benefits of a Regulatory Cutoff Level

Public Assurance. The risk of harm from exposure to radiation has been a subject of great concern to the public. It appears that most members of the public make little or no distinction between high and low doses or between a certainty of injury and a very small risk. Indeed, this lack of understanding has resulted in inordinate fears on the part of the public to such an extent that the President's Commission on the Three Mile Island accident[61] found the major health effect of the accident to have been psychological stress induced by fear and anxiety.

Many observers believe that the concept of a negligible dose, proclaimed by an expert governmental agency, will be a helpful one in educating the public to the relative unimportance of small doses of radiation. It may help, as well, in bringing to public attention the idea that linear radiation risk projections are probably conservative and that, indeed, radiation at levels around background might even be beneficial to man.

The basing of a cutoff level on comparison to background would also help educate the public to the existence of background radiation, its variability and the negative results of epidemiological studies of populations subject to relatively high background exposures.

Economic Savings. The adoption of a regulatory cutoff policy would result in economic benefits in several areas. First of all, it might eliminate the necessity for carrying out expensive epidemiological studies of effects of small doses superimposed on background levels. For example, in the Three Mile Island area, where an accidental release of radioactive material resulted in a maximum off-site dose (conservatively estimated) of 80 mrems, millions of dollars are being spent on epidemiological studies, which may be justified in public relations grounds but cannot be expected to detect any radiogenic health effects against the background of other risks. The devotion of scarce resources, money, and manpower to such scientifically futile tasks might be less likely if there were a strong, accepted, regulatory cutoff policy.

There are many analyses done to determine safety, environmental impact, or cost-benefit balance, where doses are calculated on minute levels and integrated over vast expanses of space and time. With the application of a regulatory cutoff level, this work could be truncated at a reasonable point, with no sacrifice of health and safety but with a saving in money and manpower. The same is true of ALARA calculations. Health physicists would then have more time and resources to devote to those radiation problems that may have actual public health effects. Society as a whole would be better able to devote its resources to control of more hazardous or threatening environmental agents, with an actual saving of lives. Thus, the application of such a regulatory cutoff policy is likely to result in economic savings to the government, to licensees, and to the public.

Deregulation. Recently the concept of deregulation has gained prominence. The limits of what government can do may have been reached. Because of the potential hazards that large quantities of radioactive materials of the fission and fusion processes may pose, complete deregulation of nuclear energy does not seem feasible. However, there is no need for the federal government to regulate aspects of nuclear energy that are not hazardous. The deregulation of radiation doses below reasonable cutoff levels is in keeping with the philosophy of deregulation.

Optimum Use of Resources. Studies have shown that the amount of money spent per potential death averted is much greater when the hazard is nuclear energy than when it is a more familiar one. This is not an optimal approach to the use of society's resources. Applications of a regulatory cutoff policy could tend to direct some resources now devoted to eliminating negligible risks to more gainful applications. In particular, consideration of the best use of our manpower resources indicates that the talents of toxicologists, epidemiologists, oncologists, and other specialists could be more effectively used in the control of other carcinogens, and that health physicists, radio-biologists, and nuclear engineers could better apply themselves to areas where more than negligible radiation risks are of concern.

CONCLUSIONS AND RECOMMENDATIONS

Conclusions

The following conclusions are those of the author of this chapter and are based on the information presented in the foregoing sections.

1. *Establishment of De Minimis Levels for Exposure to Radiation and Radioactive Materials Is Technically Feasible*

Some of the reasonable approaches to establishing de minimis levels based on objective scientific criteria have been discussed in this report. Several of them, including those related to background levels of dose and risk, could be used immediately without any further research or data-collection efforts.

2. *Comparison with Natural Background Radiation Levels Provides the Best Current Scientific Basis for Establishing De Minimis Levels*

A de minimis level specified in terms of radiation level or radioactive material concentration can be established by comparison to the analogous levels observed in the natural background. The spatial and temporal variability of natural background levels is such that incremental dose levels on the order of 100 mrems per year or more fall well within the envelope of background variability.

The use of background radiation levels, rather than background risk levels for comparison seems preferable because it avoids the need to adopt a specific dose–risk correlation, to make extrapolations in areas of great uncertainty, and to use *upper limits* rather than *best estimates*. The rationale is that whatever the risk level correlated with incremental radiation exposure, it is the same risk level as would be correlated with a variation in background level of similar magnitude. Therefore, as long as the incremental de minimis level is well within the envelope of background variations, it cannot pose an untoward risk to man.

3. *Establishment of Regulatory Cutoff Levels for Radiation Exposure Would Provide Benefits to Licensees and the Public That Outweigh Any Costs and Risks Involved*

The benefits include savings in the money, time, and manpower of licensees and regulators, resulting from deregulation below the cutoff level, which could then be applied to the solution of technical problems in areas where there are significant health and safety concerns. A reduction in public fears due to misunderstanding of the impacts of low level radiation could be a significant benefit. The costs of such a policy would be small since the *upper limit* estimate of the residual risks that may be involved if the policy is adopted is small and would not be detectable in the noise of background risk variability. This report has not attempted quantification of benefits and risks of a cutoff policy, but there is no reason why a reasonable estimate could not be made.

4. *At the Present Time, the Adoption of the Concept of Regulatory Cutoff Levels Is More Important than the Actual Levels Adopted*

The specific radiation dose and radioactive materials concentration cutoff levels specified can be adjusted to meet changing criteria. The important thing is that the concept of a regulatory cutoff level for radiation exposure, below which there is no regulatory concern, be recognized. Even if the regulatory cutoff level were initially set as a very small dose value (say, 1 mrem-per-year), some beneficial effect would result, since meaningless calculations involving individual doses in the micro- and picorem-per-year range could be eliminated.

5. *There Are Several Methods by Which Such a Policy Could Be Promulgated*

The best method of promulgation of a regulatory cutoff policy for radiation exposure would be a federal guidance document, prepared by EPA and signed by the president, issued as guidance for all federal agencies. However, other methods, such as adoption of such a policy by the Nuclear Regulatory Commission, or legislative enactment declaring such a national policy, would also provide a basis for the development of regulatory cutoff levels for applicability to NRC licensees.

6. *There Are a Variety of Ways That Proposals for Such a Policy Could Be Presented*

A regulatory cutoff policy could be proposed formally—for example, by petition for rulemaking addressed to NRC or EPA—or informally—by letter to, or discussion with, personnel of the agencies involved. Information on the effects of such a policy could also be presented to the news media, to state officials, and to the public. The help of professional and scientific societies concerned, as well as the standards-setting bodies, should be sought.

Recommendations

1. *Nuclear Utilities Should Support Establishment of a Regulatory Cutoff Policy with Cutoff Levels Based on Scientifically Determined De Minimis Levels*

This may involve petitioning for establishment of such policy, participating in rule-making proceedings, and other supporting activities. The EPA and NRC are the federal agencies most involved and it is recommended that participation in the proceedings of both be encouraged. It is further recommended that industry consult with cognizant scientific organizations and standard-setting bodies.

2. *Further Work Should Be Undertaken to Quantify the Benefits and Costs of Such a Policy in Specific Areas Related to Power Reactor Licensees*

This may involve the collection of information as well as analysis, modeling, and calculations.

3. *Nuclear Utilities Should Encourage Further Development of Scientific and Technical Approaches to Establishing De Minimis Levels Based on Current Radiobiological Knowledge*

For example, it may be appropriate to convene a meeting of interested scientists to discuss methodologies and data applicable to establishment of de minimis levels.

4. *Nuclear Utilities Should Support Public Education Programs Relating to the De Minimis Concept and the Policy of Regulatory Cutoff*
5. *Nuclear Utilities Should Encourage the Development of Specific Proposals for the Application of De Minimis Doses to Establishment of Regulatory Cutoff Levels in Areas of Importance to Power Reactor Licensees*

For example, drafting of proposed regulations for cutoff levels applicable to ALARA calculations in the power reactor context should be encouraged.

APPENDIX

Proposals for De Minimis Radiation Levels

Dr. Merrill Eisenbud.[1]   In a paper presented at the annual meeting of the NCRP in April 1980, Dr. Eisenbud raised the question of whether there is "a dose of ionizing radiation that is so trivial that it can serve as a cutoff point below which radiobiologists, health physicists, and regulators should ignore the caveat, 'all unnecessary radiation exposure should be avoided.'" Among the benefits of having such a level, he included reassurance to the public in the event of releases of relatively small amounts of radioactive materials and reduction in the time, effort, and space now devoted to analyses resulting in "absurdly low dose estimates," which could be better employed elsewhere.

Dr. Eisenbud discusses a variety of methods for arriving at a de minimis dose, all of which are among those discussed in the section on the technical context in this chapter. While he does not propose a specific quantitative value for such a level, he refers with apparent approval to a level of about 20 mrem per year, based on variation in natural background levels, calculated by Adler and Weinberg (see below). Apparently, he would consider a dose equivalent value up to about 100 mrem as possibly de minimis, since he states: "What the general public needs is assurance that low doses of ionizing radiation (i.e., those lower than about 100 mrem per year) will not be harmful to them in a *perceptible* way" (emphasis in original).

Dr. Harald H. Rossi.[24,25]   In two letters to the editor of *Health Physics*, Dr. Rossi discusses the need for a de minimis dose associated with the practical application of the ALARA principle that exposure to ionizing radiation be kept as low as reasonably achievable. He notes that since reducing dose rates to minute levels may be "reasonably achievable," the ALARA principle is incomplete; it does not define a stopping point for this reduction. He suggests that there be defined a dose level that is of "no concern" or "negligible," and that the ALARA principle be formulated as follows: "personal exposures [should be kept] below maximum permissible limits and reduced as near as practicable to de minimis levels. Further reductions are not required."

He proposes that a de minimis dose-equivalent index rate at any accessible location near a radiation facility, in mrems per hour, be established by "whoever demands ALARA." The major benefit discussed by Dr. Rossi for such a system is the reduction in the potential ambiguity of regulatory enforcement of the ALARA principle. As it is now formulated, there can be widely varying interpretations of what is *reasonably achievable*. According to Dr. Rossi, "the dose reduction measures decreed by one inspector may be deemed insufficient by another one."

Although he does not specifically recommend any value, as an example. Dr. Rossi refers to "a de minimis dose rate that is, say, 30% of that due to natural background."

Dr. G. Hoyt Whipple.[2]   Dr. Whipple defines "a *practical threshold* dose rate below which no observable effects on human health will be produced" (emphasis added). The major benefit of adopting such a threshold would be to avoid the "disproportionate effort . . . being expended on regulating, measuring, and recording radiation doses of no possible significance to health. This effort could be devoted to more worthwhile activities."

Dr. Whipple's approach to defining a practical threshold level is to compare the risk of low level radiation to variations in the rate of naturally occurring fatal cancers and to other statistical risk levels. In light of the cancer statistics, Dr. Whipple derives "practical threshold: levels for low-LET and high-LET radiation of 500 mrems per year and 170 mrems per year respectively, above natural background." He then suggests that a practical threshold for regulatory purposes might be "even as small as 100 mrems per year" above natural background.

Drs. G. A. M. Webb and A. S. McLean.[26]   In their report, "Insignificant Levels of Dose: A Practical Suggestion for Decision Making," Drs. Webb and McLean define "insignificant" levels of dose and risk as "low levels of dose and risk which the individual does not take into account in his decision making."

According to the authors, the benefit of such a system is that it will "codify and make consistent existing decision making criteria in relation to widely disparate practices. It will . . . [direct] attention and effort away from massive analyses of trivial doses back to those higher doses which may, indeed, be of significance and which, if possible, should be reduced."

Drs. Webb and McLean derive their insignificant dose level from consideration of the concept of a "negligible level of risk." This they define as the level of risk that is not taken into account by an individual when making decisions. They consider quantitative levels of negligible risk proposed by several authors, and select an annual risk of 10^{-6} (1 in 1 million) as one that is clearly negligible. They correlate that with a whole-body gamma dose of less than 10 mrems per year. They then select 10 mrems per year as their "insignificant" level of dose.

Drs. Webb and McLean go on to propose a system for ignoring these insignificant doses in calculating collective doses for use in practical decision making. They do so by defining a "cutoff dose," for collective dose calculations, that is a small part of the "insignificant dose level," discussed above. The cutoff level is such that it is extremely unlikely that a person exposed to many radiation "practices" would accumulate more than 10 mrem above background in a year. They propose a "cutoff" dose for a single practice of 0.1 mrem per year.

Drs. A. M. Weinberg and H. T. Adler.[27] In a letter published in *Health Physics*, Drs. Weinberg and Adler discuss a possible standard for exposure to low-dose-rate radiation, based on the suggestion that the effects of an additional dose imposed on humanity as a result of human activities would be undetectable and acceptable as long as that dose is *small* compared to natural background." They propose that "small compared to natural background" be interpreted as the standard deviation (weighted with the exposed population) of the natural background. They note that if this statistic is used as the standard (i.e., as the allowable exposure to a maximally exposed person), then the actual exposure to the population at large (average exposure) would likely be much less.

Using state-by-state data on annual doses due to natural background in the United States, they calculated a mean of 130 mrems-per-year and a standard deviation of about 20 mrems-per-year. They note that this value is comparable to the 25 mrems per year suggested by EPA as the maximum exposure to an individual in the population due to uranium fuel cycle activities. Apparently, their value is proposed as a standard, i.e., to be used as a regulatory upper limit, rather than as a de minimis level of the type described elsewhere in this chapter. However, it appears that most observers who have cited the work of Drs. Weinberg and Adler have considered this a level for application as a regulatory cutoff dose.

ACKNOWLEDGMENTS. This report was prepared for and is reprinted with the permission of the Edison Electric Institute. The author expresses thanks for the assistance and suggestions of Mr. Saul J. Harris, Edison Electric Institute; Dr. Robert W. Deutsch, General Physics Corporation; Dr. Merrill Eisenbud, New York University; and Jay E. Silberg, Esq., Shaw, Pittman, Potts and Trowbridge.

REFERENCES

1. M. Eisenbud, "The Concept of De Minimis' Dose" (NCRP meeting April 2, 1980, see reference 32).
2. G. H. Whipple, "A Practical Threshold for Radiation" (June 10, 1980).
3. S. Harris, "Radiation Levels Safe Enough to Exempt" (Unpublished manuscript, 1981).
4. NCRP Report 64, "The Influence of Dose and its Distribution in Time (Dose Rate) on Dose-Response Relationships for Low LET Radiation," NCRP (1980).
5. NUREG/CR-1174, "A Study to Determine the Feasibility of Conducting Epidemiologic Investigations of the Health Effects of Low-Level Ionizing Radiation," Phase 1, Equifax (1980).
6. V. P. Bond, "Responses to the Low-Level Radiation Controversy" (AIF conference, 4–7 October, 1981).
7. T. D. Luckey, "Hormesis with Ionizing Radiation" CRC Press, Florida (1978).
8. R. J. Hickey et al., "Low Level Ionizing Radiation and Human Mortality: Multi-Regional Epidemiological Studies," *Health Physics 40* 625–641 (1981).
9. L. Sagan "Some Thoughts on Dose-Response, Hormesis and All That," *Nuclear News*, October 1981, p. 80.
10. Los Alamos National Laboratory News Release "Rocky Flats Mortality Study Results Mean Less Worry for Plutonium Workers" (October 15, 1981).
11. R. J. Hickey, "Hormesis" (Letter), *Nuclear News*, November 1981, December 1981.
12. T. D. Luckey, "Hormesis" (Letter), *Nuclear News*, December 1981.
13. R. J. Hickey et al., "Low-Level Ionizing Radiation and Human Mortality," *Health Physics 40* 625 (1981).
14. J. W. Baum, "Population Heterogeneity and Radiation Induced Cancer," *Health Physics 40* 625 (1981).
15. E. P. Radford, "Statement Concerning the Current Versionof Cancer Risk Assessment in the Report of the BEIR III Committee" (BEIR III Report) (1981).
16. "The Effects on Populations of Exposure to Low Levels of Ionizing Radiation: 1980" (BEIR III) National Academy of Sciences.
17. H. H. Rossi, "Separate Statement—Critique of BEIR III" (BEIR III Report) (1981).
18. B. L. Cohen, "Proposals on Use of BEIR III Report in Environmental Assessments," *Health Physics 41* 769 (1981).
19. R. Evans, "Radium in Man," *Health Physics 27* 497 (1974).
20. N. A. Frigerio et al., "The Argonne Radiologic Impact Program, Part I, Carcinogenic Hazard from Low-Level Low-Rate Radiation," ANL/ES-26 Part I (September 1973).
21. U.S. Radiation Protection Council, "Report of the Task Force on Radon in Structures," RPC-80-002 (August 1980).
22. N.I.H. Publication No. 80-2087, p. 111 (1980).
23. L. Battist et al., "Population Dose and Health Impact of the Accident at the Three Mile Island Nuclear Station," NUREG-0558, (May 1979).
24. H. H. Rossi, "What are the Limits of ALARA," *Health Physics 39* 370–371 (1980).
25. H. H. Rossi, "Reply to Drs. Beninson and Lindell," *Health Physics 41* 685–686 (1981).
26. G. A. M. Webb and A. S. McLean, "Insignificant Levels of Dose: A Practical Suggestion for Decision Making," NRPD-R62, National Radiological Protection Board (April 1977).
27. H. L. Adler and A. M. Weinberg, "An Approach to Setting Radiation Standards," *Health Physics 34* 719–720 (1978).
28. A. M. Weinberg, "Energy Policy and Mathematics," ORAU-IEA-79-8 (0) (1979).
29. "Perceptions of Risk," Proceedings of the 15th Annual Meeting of NCRP, March 1979 (CRP Proceedings No. 1).
30. "Recommendations of the International Commission on Radiological Protection," ICRP Publication 26 (1977).
31. "Implications of Commission Recommendations That Doses Be Kept as Low as Readily Achievable," ICRP Report 22 (1973).
32. "Quantitative Risk in Standards Setting," Proceedings of the 16th Annual Meeting of NCRP, April 1980 (NCRP Proceedings No.2).
33. "Report to the American Physical Society by the Study Group on Nuclear Fuel Cycles and Waste Management," 50 Reviews of Modern Physics, No. 1, Part II (January 1978).
34. National Research Council, Advisory Committee on the Biological Effects of Ionizing Radiation, "Effects on Populations of Exposure to Low Levels of Ionizing Radiation," Washington, D.C., BEIR I (1972).

35. Public Law 830703 "The Atomic Energy Act of 1954," as amended (42 U.S.C. 2011, et seq.).
36. Public Law 95-604 "Uranium Mill Tailings Radiation Control Act of 1978" (42 U.S.C. 7901).
37. Public Law 91-190 "National Environmental Policy Act of 1979."
38. In the Matter of Uranium Mill Licensing Requirements, CLI-81-9, 13 NRC 460 (1981).
39. NUREG-0757, "Radon Releases from Uranium Mining and Milling and Their Calculated Health Effects," U.S. N.R.C. (February 1981).
40. NUREG-0706, "Final Generic Environmental Impact Statement on Uranium Milling," USNRC (September 1980).
41. Kerr-McGee Nuclear Corp. vs. NRC, CA NO. 80-2043 (9th Cir., June 1981), Brief Amicus Curiae of Mountain States Legal Foundation.
42. LBP-78-25, 8NRC87 (July 14, 1978).
43. In the Matter of Philadelphia Electric Co., et al., ALAB- 640 13NRC 487 (1981).
44. ALAB-640 Dissent, citing ALAB-5098, 8 NRC at 684.
45. ALAB-175 (1974); ALAB-161 (1973).
46. NRC, "Disposal of Low-Level Radioactive Wastes; Availability of Preliminary Regulations" (10CFR Part 61), *Federal Register 45* 13105–13106 (February 28, 1980).
47. U.S. Radiation Policy Council "Progress Report and Preliminary 1981-83 Agenda" (RPC-80-001) (September 30, 1980).
48. "Biomedical Waste Disposal," Amendments to 10CFR20, *Federal Register 46* 16230, (March 11, 1981).
49. "Value/Impact Statement of Amendments to 10CFR20 for Disposal of Biomedical Wastes," *U.S. NRC (March 2, 1981)*.
50. *Federal Register 45*, pp. 71761–71762, 1980.
51. "Disposal or On-Site Storage of Thorium or Uranium Wastes from Past Operations" (Notice for Comment) *Federal Register 46* 52061 (October 23, 1981).
52. NUREG/CR-2082, "Monitoring for Compliance with Decommissioning Termination Survey Criteria" (ORNL/HASRD- 95) (1981).
53. Communication from S. Harris, EEI.
54. W. A. Rodger et al., "De Minimus (*sic*) Concentrations of Radionuclides in Solid Wastes," prepared for AIF NESP, (AIF/NESP-016) (April 1978).
55. C. Comar, "Risk: A Pragmatic De Minimis Approach" *Science 203* 4378 (January 26, 1979).
56. C. A. Willis and F. J. Congel, "Status of NRC Radiological Effluent Technical Specification Activities" (AIF Conference 4–7 October, 1981).
57. Peshlakai v. Duncan, 476 F. Supp. 1247 (D.D.C. 1979).
58. G. Voelz and A. Valentine, "The Karen Silkwood Case," ANS Meeting November 1981. On December 11, 1981, a panel of the Court of Appeals for the 10th Circuit rejected the jury award of damages for injuries and said the claim should have been brought under State Workers Compensation Law.
59. See NUREG-0739.
60. Concurring and Dissenting Opinions in "Benzene Case," Industrial Union Department, AFL-CIO vs. American Petroleum Institute, 442 US 938.
61. Report of the President's Commission on the Three Mile Island Accident (1979).
62. NUREG-0436, Rev. 1, "Plan for Reevaluation of NRC Policy on Decommissioning of Nuclear Facilites" (1978).
63. Regulatory Guide 1 86, "Termination of Operating Licenses for Nuclear Reactors," USNRC (June 1974).
64. Report of the Inter-Agency Task Force on Compensation for Radiation-Related Illnesses (February 1, 1981).

BIBLIOGRAPHY

"The Applications of Cost-Benefit Analysis to the Radiological Protection of the Public," National Radiological Protection Board (UK) (March 1980).

Brookhaven National Laboratory, "Low Level Radiation Effects—Fact Book" (Draft).

10CFR51 Appendix A, "Narrative Explanation of Table S-3, Uranium Fuel Cycle Environmental Data" (proposed rule)*Federal Register 46* 15154–15157 (March 4, 1981).

J. J. Cohen, "Indicators of Possible Beneficial Effects of Low-Level Radiation," ANS Meeting (ANS Transactions) (November 1981).

Department of Energy Regulations 10CFR712, "Grand Junction Remedial Action Criteria."

DOE/EIS-0046F, Final Environmental Impact Statement, "Management of Commercially Generated Radioactive Waste," U.S. Dept. of Energy (October 1980).

EPA, "Part 190—Environmental Radiation Standards for Nuclear Power Operations," *Federal Register 42* 2858 (January 13, 1977).

EPA, "Part 192—Environmental Protection Standards for Uranium Mill Tailings," *Federal Register 42 27367* (April 22, 1980).

EPA, "Persons Exposed to Transuranium Elements in the Environment," Proposed Federal Radiation Protection Guidance on Dose Limits, *Federal Register 42* 60956 (November 30, 1977).

EPA, "Proposed Disposal Standards for Inactive Uranium Processing Sites; Proposed Rule and Extension of Comment Period," *Federal Register 46* 2556 (January 9, 1981).

"Estimates of Ionizing Radiation doses in the United States 1960–2000," ORP/CSD 72-1, U.S. EPA.

"Guidelines Pertinent to the Development of Decommissioning Criteria for Sites Contaminated with Radioactive Material" (August 1978).

"Impact Statement on 10CFR Part 61 Licensing Requirements for Land Disposal of Radioactive Wastes," U.S.N.R.C. (September 1981).

In Re Philadelphia Electric Company, Docket No. 50-277, U.S. Nuclear Regulatory Commission, "Licensees' Response to NRC Staff Motion for Leave to Include Health Effects Findings" (July 18, 1980).

"Licensing Requirements for Land Disposal of Radioactive Waste," *Federal Register 46* 51776 (October 22, 1981); *Federal Register 46* 38081 (July 24, 1981) (Proposed Regulations).

B. Lindell and D. Beninson "ALARA Defines Its Own Limit," *Health Physics 41* 685–686 (1981).

"Low Level Ionizing Radiation," Hearings before the House Committee in Science and Technology, 96th Congress, 1st Session (June 1979).

NCRP Report No. 43, "Review of the Current State of Radiation Protection Philosophy" (January, 1975).

NCRP Report No. 45, "Natural Background in the United States" (1975).

NCRP Report No. 50, "Environmental Radiation Measurements" (December 1976).

NRC, "Exemption of Technetium-99 and Low-Enriched Uranium vs. Residual Contamination in Smelted Alloys," 10 CFR Parts 30, 32, 70, and 150, Proposed Rule, *Federal Register 45* 70874 (October 27, 1980).

NUREG-0586, "Summary of Preliminary Draft Generic Environmental Impact Statement on Decommissioning Nuclear Facilities" (Preliminary Draft).

NUREG-0613, "Residual Radioactivity Limits for Decommissioning" (Draft Report) (1979).

NUREG-0782 Draft Environmental Impact Statement on 10CFR Part 61, "Licensing Requirements for Land Disposal of Radioactive Wastes;" Volume 2 – Main Report; Volume 4 – Appendix N, "Analysis of Existing Recommendations, Regulations and Guides" (1981).

NUREG/CR-0573, C. C. Travis et al., "A Radiological Assessment of Radon-222 Released from Uranium Mills and Other Natural and Technology Enhanced Sources."

NUREG/CR-0671, A. H. Schilling et al., "Decommissioning Commercial Nuclear Facilities: A Review and Analysis of Current Regulations" (1979).

NUREG/CR-2226, L. Lave and T. Romer, "A Survey of Safety Levels in Federal Regulation" (1981).

D. Oakley, "Natural Radiation Exposure in the United States," ORP SID 72-1 (EPA).

"Radiological Quality of the Environment in the United States, 1977" (EPA 520/1-77-009).

P. F. Ricci & L. S. Molton, "Risk and Benefit in Environmental Law," *Science 214* 1096 (December 4, 1981).

P. Slovic et al., "Informing the Public About the Risks from Ionizing Radiation" *Health Physics 41* 589–598 (1981).

U.S. Nuclear Regulatory Commission, "Applicant's Memorandum on Radon Emissions and in Support of the De Minimis Approach" (April 9, 1979).

"Use of Uranium Mill Tailings for Construction Purposes," Hearings before the Joint Committee on Atomic Energy, 92nd Congress, 1st Session (1971).

Index